DeColonize EcoModernism!

ALSO AVAILABLE FROM BLOOMSBURY

Philosophical Posthumanism, Francesca Ferrando
Why Climate Breakdown Matters, Rupert Read
General Ecology, Erich Hörl
Ecofeminism as Politics, Ariel Salleh

DeColonize EcoModernism!

Ariel Salleh

The Androcene and its Others – Vol I

BLOOMSBURY ACADEMIC
LONDON • NEW YORK • OXFORD • NEW DELHI • SYDNEY

BLOOMSBURY ACADEMIC
Bloomsbury Publishing Plc
50 Bedford Square, London, WC1B 3DP, UK
1385 Broadway, New York, NY 10018, USA
29 Earlsfort Terrace, Dublin 2, Ireland

BLOOMSBURY, BLOOMSBURY ACADEMIC and the Diana logo are trademarks of
Bloomsbury Publishing Plc

First published in Great Britain 2024

Copyright © Ariel Salleh, 2024

Ariel Salleh has asserted her right under the Copyright, Designs and Patents Act, 1988, to
be identified as Author of this work.

For legal purposes the Acknowledgements on p. xviii constitute an extension of
this copyright page.

Series design: Charlotte Willow Retief

All rights reserved. No part of this publication may be reproduced or transmitted
in any form or by any means, electronic or mechanical, including photocopying,
recording, or any information storage or retrieval system, without prior permission in
writing from the publishers.

Bloomsbury Publishing Plc does not have any control over, or responsibility for, any
third-party websites referred to or in this book. All internet addresses given in this
book were correct at the time of going to press. The author and publisher regret
any inconvenience caused if addresses have changed or sites have ceased to exist,
but can accept no responsibility for any such changes.

A catalogue record for this book is available from the British Library.

A catalog record for this book is available from the Library of Congress.

ISBN: HB: 978-1-4742-7761-7
PB: 978-1-4742-7760-0
ePDF: 978-1-4742-7762-4
eBook: 978-1-4742-7765-5

Typeset by Deanta Global Publishing Services Chennai India
Printed and bound in Great Britain

To find out more about our authors and books visit www.bloomsbury.com
and sign up for our newsletters.

To Lau Kin Chi and Sit Tsui
in awe of your inspirational energies
in building the
Global University for Sustainability

Contents

Preface to the Trilogy A Subliminal Reading of Political Ecology xi
Acknowledgements xviii

1 Resisting Extinction: *Youth Join the Dots* 1
   Anthropocene 2
   Androcene 4
   The Master Imaginary 8
   Patriarchal-colonial-capitalism 10
   Global Debt Matrix 12

2 Global Synergies: *Livelihoods or Lifestyles?* 19
   The Activist 20
   The Teacher 23

3 *Terra Nullius: Consuming Lands and Bodies* 29
   Extractivism 30
   Exterminism 32
   BioColonialism 35
   Nonsensical Law 38

4 Nuclear Risks: *Voices for Life-on-Earth* 43
   Denialism 44
   Women's Collectives 46
   From Victims to Leaders 50
   Enough Looting! 52
   Postscript 55

5 Earth System Governance: *Uncertainty Principle Revisited* 57
   Conceptual Fit 60
   EcoModernism 63
   Reducing Complexity 66
   Steering Laissez-faire! 68
   Validity 71

6 The Gene Trade: *Organized Irresponsibility* 75
   Measurable Units 76
   Unpredictable Risk 78
   Matters Outstanding 80
   Coexistence 81
   Andro Science 83

7 *Buen Vivir*: *Ecomodernist or Andean?* 89
   Strategy I 89
   Embodied Debt 91
   Strategy II 94
   Back to Dependency 97
   Metabolic Value 100
   Eco-Sufficiency 102

8 Climate Science and Water: *Coming to Our Senses* 107
   Carbon Fetishism 108
   Methodological Forcing 110
   Scale Versus Responsibility 113
   Peoples' Science 115
   Moving On 117

9 A Just Transition?: *Women Are the Key* 119
   Andro Alienations 120
   Embodied Thinking 121
   Global Actors 123
   Regenerative Models 124

10 Food Sovereignty: *Meeting Real Needs* 127
  Internal Colonies 128
  Benefit Sharing 131
  Whose Capacity? 133
  Meta-Industrial Labour 136
  Agri-tech 138

11 Another Future Is Possible!: *Holding Ground* 141
  Andro Others 142
  Major Groups 145
  Coloniality 150
  A Biocivilization 152
  Real Green Jobs 154

12 Green New Deals: *For Globalization Lite* 157
  UK and UNEP 158
  Transatlanic 161
  Australia 164
  US Democrats 167
  EU and DiEM25 168
  DSA-USA 170
  Others' Deals 172

13 The 2030 Agenda: *Sustainable Development Goals* 175
  Fixing Poverty 176
  High-Tech Designs 178
  Climate Finance 180
  False Consensus 182
  Water Is Life 184

14 The Smart ReSet: *A Biopolitical Turn* 189
  The Fourth IR 190
  Internet of Things 193
  Captured Agencies 197

New Climate Impacts 200
Colonizing Space 203

15 Digital Coloniality: *Everyday Contradictions* 209
Modernity 212
Data Sovereignty 214
Double Binds 217
Decoupling? 219

16 Land Is Law/Lore: *Another Ontology* 223
Eco-Sufficient Ethics 224
Bioregional Politics 227

Notes 231
Index 271
  Author Index 271
  Subject Index 279

Preface to the Trilogy
A Subliminal Reading of Political Ecology

The Androcene and Its Others is a trilogy that argues for treating the patriarchal-colonial-capitalist system as a single political entity. This invites a shared strategy of resistance among feminist, decolonial, socialist and ecological movements. In examining how workers, women and indigenous peoples are each manipulated by a culture of systemic dualisms, the books delve into the deep structure of political ecology.

The *Androcene* is a vibrant historically evolved complex of patriarchal-colonial-capitalist privilege and rulemaking. Over millennia, it has expanded across the world from local civilizations and religious rites through tribal invasions and imperial adventures, arriving at its modern capitalist phase some 500 years ago. Intellectual landmarks in this largely Eurocentric trajectory include the classical logic of Greece, the European scientific revolution, a Hobbesian theory of the state and nineteenth-century evolutionism. However, this trilogy focuses on where the action is right now – that is, with grassroots decolonial, ecosocialist and ecofeminist movements for political change. Whereas Anthropocene scientists emphasize deep geological time, social life is infused with another time, a living metabolism of materially embodied energies – libidinal and unconscious, intentional and judgemental. The Earthwide changes that are classified under 'deep geological time' have their source in 'deep affective time'. This historical process is preconscious and clouded in denial, but its sublimated form shows up in the universalizing dualisms of Humanity over Nature, Masculine over Feminine, White over Black, Subject over Object and so on. Even in modernity, this pseudo-logic rationalizes and helps hold together the entanglement of patriarchal-colonial-capitalist institutions.

Sociology of Knowledge

Many essays in these three books – *DeColonize EcoModernism!*, *EnGender EcoSocialism!* and *ReGround EcoFeminism!* – were written in action. But they are also grounded in a wide transdisciplinary reading beyond political ecology, as we know it. If the twenty-first-century polycrisis involves a biogeochemical undoing of the planet, equally the crisis has social dimensions seen in the horrors of instrumental rationality, Global South extractivism, violence on sex/gendered and racialized bodies, the hegemony of Green Deals and endless war. Again, the indifference to life itself occurs in philosophies from structuralism to deep ecology to posthuman fantasy. On the other hand, joined-up thinking allows the multiplicity of patriarchal, colonial and capitalist practices to be understood as a single system. Each political layer in this continuum has grown out of the one before, but all three interact in an elective affinity, as sociologist Max Weber might have said. Patriarchal domination is the oldest frame and originary energizing force of the *Androcene* and remains the common denominator of global politics today. Thus, while Karl Marx judged human alienation from nature to be an expression of capitalist economics and its break with the organic conditions of life, this is only partly true. Alienations arising from the ancient Humanity versus Nature divide and its dualisms are far more fundamental and no less material in effect than the deprivations attributed to capital. Thus, in response to the Left emphasis on 'relations of production', an ecofeminist analysis demands attention to the 'relations of reproduction' that sustain modern globalization at its base. Healing the crisis of Life-on-Earth will call for much more than socio-economic and technical fixes.

The Androcene and Its Others can be read as an ecofeminist sociology of knowledge, bringing home to readers just how power and knowledge making are interwoven. For the genealogy of ecopolitics, to borrow Michel Foucault's term, has several misleading models in use, and activists need to be aware of this as they commit to the defence of Life-on-Earth. Even categories like 'time and space' are not *a priori* truths. What Emile Durkheim labelled collective representations, peoples' ways of making sense of their world, are geographically unique and set in place over time by repetition. Later, Peter Berger and Thomas Luckmann would spell out the phenomenology of just how everyday social construction lays down what passes for common sense. The Marxist dialectic likewise treats action and thought as co-created. Political ecologists are not necessarily familiar with the preconscious processes that undergird their analyses; so in taking on patriarchal-colonial-capitalist relations as a unity, activists today function rather like the free-floating intellectuals of Karl Mannheim's first sociology of knowledge. They refuse the disciplinary gaze of modern narcissism; although that said, far too many cadres still operate from inside the master's house.

Androcentrism is already clear in the ancient symbolism of threatening mother figures – the mythologies of Medea or Kali, for example, not to mention an unruly Mother Nature herself. Mark Furlong writes,

> Deep in the psychic interior, these dark tropes survive as vestigial, embedded images . . . Confronted by climate collapse, bourgeoise psychology and neoliberal ideology will jointly advocate that the self be militarised – that personal boundaries be strengthened, feelings cauterised, behaviours made strategic . . . Misogyny cannot understand that everything that lives exists interdependently. Nothing stands, or falls, on its own . . . Recognition of climate collapse is not only about the quantum and distribution of information. A complex psychical precondition must also be understood.[1]

An ecofeminist sociology refuses the ideologies of patriarchal-colonial-capitalist power. Since Plato, the Eurocentric mindset armed itself with universal and timeless essences, precursors of the contemporary obsession with measurable abstractions. Now the rise of new digital technologies leans heavily on the Cartesian vision of an objectified external world regulated by mathematical laws. But a culture that separates Humanity from Nature functions by denial, its split consciousness relying on arbitrary dualisms to make sense of things. As Andrea Saltelli and Monica Di Fiore point out, scientific modelling too easily paves over the intersection of psychological rationalization and technical realities.² Then again, recent work in the Humanities has favoured static totalizing synchronic models based on discourse analysis, leaving a gap in the explanation of how social life is constituted over time. An adequate sociology will draw empirically on peoples' experience, including intentional and preconscious motivations. This is where activism can begin to sharpen theory. Consider the worldwide resistance known as the 'movement of movements', a praxis that links the concerns of women, indigenous peoples, workers and youth. As a transversal politics based on an 'embodied materialism', it can restore peoples' common cause with each other and with the Earth. In turn, it opens a way into 'eco-centric thinking'. In Furlong's words: 'wholeness requires that the anthropogenic world view be decentered.'

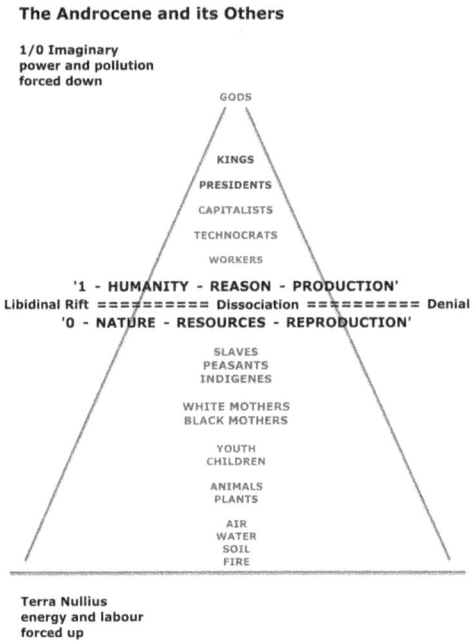

The Androcene and Its Others: A Debt Matrix

DeColonize EcoModernism!

In the twenty-first century, the old colonial attitude of *terra nullius*, meaning a vacant place, free for the taking, still lurks behind the global economic expropriation of peoples' lands and bodies; but today, the theft is rationalized by ecomodernist policy. This book enters into that *Androcene* – its climate politics and nuclear risks, mining and the gene trade, the Fourth Industrial Revolution and digital coloniality. It spells out the social and ecological contradictions set in motion by contemporary neocolonialism. The patriarchal-colonial-capitalist imperium and sometimes even environmental activists themselves advocate Green New Deals, Earth Governance, Sustainable Development Goals and Smart Futures. Meanwhile, the word from decolonial thinkers like Arturo Escobar in Colombia and Tyson Yunkaporta in Apalech country, or feminist technology critics like Vandana Shiva in India and Shoshana Zuboff in the United States, is that the dispossession of First Nation peoples' livelihoods is not repaired by giving them access to lifestyle consumerism. Worldwide social movement activists readily see through the *1/0 imaginary* and its charade of Earth Summits. Youth, especially, is standing up to the ruling class and its extinction trajectory. So too, those who are ecologically aware refuse the fashionable posthumanist ideology that circulates in high-tech quarters. Beyond 'exchange value', the Others of the *Androcene* are building a bioregional future, respectful of indigenous skills; they want food sovereign economies, protective of nature's 'metabolic value'. These self-governing models include *buen vivir*, ecovillages, *swaraj* and commoning.

EnGender EcoSocialism!

Systemic critiques of capital accumulation from Left scholars like Ulrich Brand or Jason Hickel are reminders of how global production chains are subsidized by the labours and lands of nameless Others. The inhabitants of this *terra nullius* are politically invisible, but they include mothers who birth workers and freely provide domestic care as well as forest dwellers and small holders skilled in regenerative provisioning. But ecosocialism has been slow to grasp the class basis and ecological significance of womens and indigenous 'meta-industrial labours'. From an embodied materialist perspective, the epistemological emphasis on structure versus consciousness in Left analysis appears to be another exemplar of the Humanity versus Nature dissociation.

Ecosocialists underscore the 'metabolic rift' resulting from capitalist over-resourcing of rural ecologies, but they overlook the *libidinal rift* or sex/gender blind spot in their own political theory. This book journeys through James O'Connor's 'second contradiction thesis', John Bellamy Foster's 'metabolic rift', Joel Kovel's theory of 'ego', Jason Moore's 'four cheaps' and Kohei Saito's recent translations of Marx's late work. While women's labours – domestic and theoretical – are still marginal in ecosocialist thinking, the ecofeminist notion of a 'meta-industrial class' could lead it towards an ecocentric perspective on value.

ReGround EcoFeminism!

Often feminism has been misunderstood because it has several paradigms – radical, cultural, anarcha, liberal, socialist, postmodern, transhuman, decolonial and ecological – each with different political implications. This book addresses the last of these, honouring women from Appalachia to Finland to Ecuador and beyond, showing how their labours 'hold' the materiality of Life-on-Earth together. If Marxists naturalize 'species being', decolonial realists like Rita Segato point to 'the mandate of masculinity' as culturally learned. And this mandate is not unrelated to the erasure of women's voices as played out among Deep Ecological philosophers; in the ambivalence of Social Ecology; at major UN Summits and even in Green Party formation. Since by the *libidinal rift* women are defined as 'closer to nature than men' along with 'natives and animals', many feminists have been shy of 'the nature question', fearing it might undermine their claim to equality. Donna Haraway famously tried to avoid the violence of 'patriarchal naturalization' only to end up inside the *1/0 imaginary* of technoscience. The present ecofeminist argument reconciles psychoanalytic radicals, such as Luce Irigaray and Julia Kristeva, with the grounded materialist ecofeminism of Hilkka Pietila, Maria Mies and Shannon Bell. In doing so, it argues that, paradoxically, the anguish of non-identity can energize political insight.

Biopolitics and Transversality

Notwithstanding the patriarchal-colonial-capitalist hierarchy of oppressions, life-affirming ecological futures are being nurtured by decolonial, ecosocialist and ecofeminist activists. Each grassroots voice has its place in an integrated global 'movement of movements', but it is important to know when one or other political lens can be used effectively and when not. Hopefully, the *Androcene* trilogy will be useful for this learning. Modernity has become a contest between Power-Technology-Value and Land-Labour-Body – or to use Marxist terminology, a contest between relations of production and relations of reproduction. Yet in an emerging biopolitical era, there is an important caveat to note because struggles for Life-on-Earth may destabilize old Eurocentric certitudes over what is Left and what is Right. Since Covid-19, this shift has already brought political disorientation to some committed scholars and activists. So in teasing out the entanglements of patriarchal-colonial-capitalist power, let intellectual flexibility, patience and generosity be our revolutionary virtues.

Acknowledgements

Young friends have sometimes said to me, 'you are a walking archive', and in a way that is true. Generation of '68, I came to political awareness in Tasmania, as a student raising two small Malay daughters, so the path into Left, decolonial, feminist and ecopolitics was set early. A stint at commune living taught me that while such alternatives can advance environmental justice, sex/gender democracy is another matter! A first paid job involved urban accomodation needs for transient indigenous people. Later, I taught university courses in critical sociology, gender studies, and social ecology and worked with new journals like *Thesis Eleven*, *Capitalism Nature Socialism* and *Organization & Environment*. But my real political education came with initiatives like Science for People, the Movement Against Uranium Mining, The Greens, Franklin No Dams Campaign, Engineers for Social Responsibility, Wombarra Tunnel Busters, *Planeta Femea* at Rio Earth Summit and a Government Gene Technology Ethics Committee.

The transversal political ecology proposed in this trilogy – *The Androcene and Its Others* – has been inspired by working with committed people North and South – not least, my materialist ecofeminist comrades, the late Maria Mies, Vandana Shiva, Mary Mellor, Ana Isla and Sister Mary John Mananzan. The insights of first-generation ecofeminists were often marginalized in their time but accepted now, as younger scholar-activists reframe modernity as 'a crisis of reproduction'. The grassroots resistance of Women and Life-on-Earth has given way to global leadership by the World March of Women. Positive social changes are happening, and yes, Another World Is Possible!

The early 2000s brought an invitation to join Frieder Otto Wolf's Brussels research team looking at European Union Sustainability Strategy and, later, at the Green New Deal with the Rosa Luxemburg Stiftung. This fabulous learning experience, in turn, led me into Slovakia and a chance to study Michal Kravcik's pioneering Water Paradigm. Rolf Czeskleba-Dupont at Roskilde got me working on the UN Sustainable Development Goals, and Alf Hornborg at Lund University energized my thinking on the Anthropocene. A GDR fellowship in Post-Growth Societies, hosted by Friedrich Schiller University Jena, allowed time for reading deeply on Earth System Governance. I thank the Jena sociologists warmly, especially Brigitte Aulenbacher, subsequently co-editor of the International Sociological Association newsletter *Global Dialogue*. Other supportive European comrades have been Marxist feminists Frigga Haug, Diana Mulinari and Nora Rathzel as well as Haris Golemis with Transform!, Uli Brand with RLS and Clive Spash as editor of *Environmental Values*. Another big thank you is owed to my fearless translators Rok Kranzc, Goran Durdevic and Pyotr Kondrashov, in Ljubljana, Zagreb and Ekaterinburg, respectively. In France ecofeminism is being reincarnated by Emilie Hache, Daria de Beauvais, Barbara Glowczewski, Genevieve Pruvost, Catherine Larrere and the Sorbonne Colloquium on Women, Ecology and Commitment. It has also been a thrill to find Italian associates Luigi Pellizzoni, Emanuele Leonardi, Salvo Torre and Alice dal Gobbo bringing ecofeminist ideas to the Italian Fridays for Future movement. Meanwhile, a new generation ecofeminist thinker, Stefania Barca in Spain, is at work integrating our politics with ecosocialism.

As a member of the Hong Kong-based Global University for Sustainability, I remain in awe of coordinator Lau Kin Chi – and on mainland China, the rural regeneration work of Sit Tsui and Yiching Song. I have been lucky enough to work with ecosocialist Qingzhi Huan at Peking University – and in Japan, with Shuji Ozeki and editors of *Journal of Environmental Thought and Education*. In South America, Miriam Nobre, Paulo Martins and Felipe Milanez have linked me to happenings in Brazil. A partnership with Ecuador has come through Miriam Lang and Ivonne Yanez of the Rosa Luxemburg Stiftung Working Group Beyond Development. In South Africa, the visiting professorship at Nelson Mandela University would evolve into a family of environmental philosophers, Inge Konik, Adrian Konik and Bert Olivier – reinforced from Johannesburg by activist David Fig, ecofeminist Samantha Hargreaves, ecosocialists Patrick Bond, Vishwas Satgar and Awande Buthelezi.

At the University of the Free State in Bloemfontein, I send thanks to Johann Rossouw and his team of critical philosophers for their recent hospitality.

In the United States, since the passing of Joel Kovel, fellow *Capitalism Nature Socialism* editor Saed Engel-diMauro and now Danny Faber and Leigh Brownhill are carrying the journal's vision forward – not to forget the endlessly creative Quincy Saul of *Ecosocialist Horizons*. In New York, the late Lukacs scholar Agnes Heller, an associate from *Thesis Eleven* days, and dialectician Bertell Ollman at NYU were each supportive of my early efforts in critical theory. Other significant American mentors have been Jackie Smith at *Journal of World-Systems Research* and John Bellamy Foster at *Monthly Review*. Thanks also go to Michael Bell, Michael Carolan, Julie Keller and Kathryn Legun for inviting me to write on nature in the *Cambridge Handbook of Environmental Sociology*. In the United Kingdom, appreciation is due to Elena Baglioni and Liam Campling, for a rewarding term with them at the Centre on Labour, Sustainability and Global Production, Queen Mary University of London.

Over the past decade, Political Economy at the University of Sydney has been my academic home in Australia. And here I acknowledge Frank Stilwell at the *Journal of Australian Political Economy*; Adam Morton's generous leadership; and long-time climate warrior Stuart Rosewarne. Friends Hamed Hosseini, James Goodman, Sara Motta and Barry Gills deserve acclaim for sustaining transversal political critiques across several media. My own writing has been recharged by posts from enthusiastic peace worker Frank Hutchinson; Duane Norris, coordinator of the Independent Council for Ecosystem Restoration; and EMR researchers Victor Leach, Don Maisch and S. T. In Melbourne, the *Arena* team and Institute for Post Colonial Studies, Simplicity Institute, ecosocialist Arran Gare, ecofeminist Lara Stevens, ethicist Phillip Payne and Bob Phelps of the GeneEthics network continue to provide bearings for my work.

Finally, thanks go to Ashish Kothari, Arturo Escobar, Federico Demaria and Alberto Acosta, collaborators in the *Pluriverse* (2019) project, as well as my other post-development teachers – Nawal Ammar, Janis Birkeland, Michelle Boulous Walker, George Caffentzis, John Clark, Silvia Federici, Motoi Fuse, Kirk Huffman, Renate Klein, Terry Leahy, Rabbi Michael Lerner, Michael Lowy, Father Sean McDonagh, Arvind Narrain, Luke Novak, Anne Poelina,

Jan Pokorny, John Seed, Christelle Terreblanche, Ted Trainer and Simone Woerer. In Barcelona, Joan Martinez-Alier kindly hosted our *Pluriverse* editorial consultation. Aram Ziai in Kassel, Michael Lowy in Paris, Jason Hickel and Patrick Curry in London and locally Jon Altman, Rachel Evans and Sam Alexander, each helped to launch the book. The three years given over to the *Pluriverse* project explains why the present book is so behind time!

Many ideas in the *The Androcene* series were developed for talks, lectures and workshops here and there – ABC Radio National Talks; CADTM Summer School in Namur; Catholic University of Lisbon; Chinese Academy of Science, Beijing; Degrowth Summer School, Barcelona; Institute for Political Ecology, Zagreb; Polish Biennale 2019; Summer School in Cerbere; Rosa Luxemburg Stiftung, Berlin; Roskilde University; Simon Bolivar University, Quito; a WoMin seminar in Johannesburg; and a recent World March of Women panel in Ankara.

While some chapters are new, others build on early versions of pieces that would later be published by Routledge, Cambridge University Press, Guilford, Merlin Press, Zed Books, Pluto Press and Angus & Robertson. Additionally, my thanks go to the following journals for the freedom to adapt my work for the trilogy: *Arena Magazine; Capitalism Nature Socialism; Chain Reaction; Ecuador Debate; Environmental Ethics; Environmental Politics; Futures; Globalizations; Institute for Global Development UNSW; International Critical Thought; International Journal of Water; Journal of Australian Political Economy; Journal of Environmental Thought and Education; Journal of World-Systems Research; New Leaves; New Matilda; Organization & Environment; Polygraph; Progress in Political Economy; Studies in the Humanities; The Commoner; The Ecofeminist Newsletter; The New Significance; The International Handbook of Political Ecology; Thesis Eleven;* and *ZNet*.

Affirmative reviews of the initial proposal were gratefully received from Bill Carroll, John Clark, Wendy Harcourt, Jakob Horstmann and the late Peter Waterman. And yes: a special three cheers goes to my editorial team at Bloomsbury Publishers, London, and to Matt Egan in Sydney for his magical technical support.

1

Resisting Extinction
Youth Join the Dots

As I begin this book, at home in Gadigal Country, lightning strikes set hectares smouldering; black skies; electrical flashes; chaotic winds; roaring walls of flame; thick brown smoke; charred trees; crops and livestock taken; countless wild species lost. In terror, people cower together on a beach; homes flattened; cars melting; no fuel; phones down; power out. Australia's middle-class climate refugees can thank their governments for this 'new normal' – for federal and state politicians, Liberal and Labor, each are beholden to Big Coal and gas, economic growth the first premise of policy. Indigenous First Nation peoples protected this land for over 60,000 years. Now the Australian continent's major river system, the Murray-Darling Basin, has become a dry ditch after years of water theft by corporate cotton irrigators: dying fish float to the surface in shallow pools; algae flourish; plastic bottles of drinking water are trucked in for local communities. Some months later, it will be relentless rains and flooded landscapes; 50,000 families homeless. A relentless pattern of wildfires, floods and typhoons is moving across the planet, continent by continent in turn. Is this what the Anthropocene means?

Anthropocene

Not unrelated to this is the Pentagon's hypothetical review of future climate impacts that stem from the United States' total social dependency on the power grid. In these early days of what is called the Fourth Industrial Revolution, the official Defense Agency assessment is that if power infrastructure collapses, not only will the world's greatest military machine go down but also Americans will experience

- Loss of perishable foods and medications
- Loss of water and wastewater management
- Loss of electric light, heating and air conditioning
- Loss of computer/phone, communications, satellite networks, GPS and flights
- Loss of fuel distribution and public transportation systems
- Loss of all systems that do not have backup power

Something is badly awry, when everyday living is designed in such a precarious way.[1]

The twenty-first-century approach to organizing human society may reveal ingenuity, but is it thoughtful? Wild fire outbreaks have become even more extensive and extreme, particularly along the Pacific coast of Canada and the United States, as well as in Siberia, Portugal, France and Greece. Indigenous peoples lose their cultural hunting and fishing grounds. At the same time, disturbed atmospheric weather patterns have washed away towns and farmlands in Northern Hemisphere countries from Germany to Pakistan. The ecological and social costs of high-tech globalization are clear, but those who benefit most from capitalist modernity, ruling-class corporate entrepreneurs and technocrat professionals, mostly ignore the message. Meanwhile, what happens to the others? Workers reliant on the productivist system for a living wage find themselves with few choices. Agitation grows among indigenous and peasant peoples whose livelihoods are threatened by resource extractivism; agitation grows among women trying to keep households together; and among youth making their way into a risky future. After decades of privatization and deregulation by neoliberal governments, indifferent to everyday communities and their needs, bureaucratic responses to the Covid-19 pandemic would make inequalities worse – citizens taking to the streets in frustration. In an editorial for the journal *Globalizations*, Barry Gills talks of a conjuncture of crises with 'all the cumulative entropic tendencies of historical time being combined and compressed into the present. The world as we know it is literally breaking down.'[2] A now familiar rehearsal of 'failed states' lists Chile, Thailand, Nigeria, Georgia and India – and some media commentators even add post-Brexit UK, as well as the United States post-Trump. The Agrarian South Network asks, is revolution inevitable?[3]

A marked sociological divide exists between beneficiaries of this Anthropocene era and those whose lives are destroyed by industrial development, militarized land grabs and vacillating climate policy. Building on the Rio Earth Summit, a process ongoing since 1992, the business sector, politicians, World Bank and United Nations Environment Programme pursue various sustainability propositions. But this patriarchal-colonial-capitalist imperium would protect Life-on-Earth by commodifying and marketing the 'ecosystem services' already freely given in natural material flows between sunlight, forests, rivers and soils. The International Monetary Fund advances a so-called Green Economy based on pricing mechanisms, as do corporate-captured lobby groups like Conservation International, Worldwide Fund for Nature, the Environmental Defense Fund and Nature Conservancy. Routine decision-making by world leaders relies on a confused rhetoric of 'financial capital', 'human capital', 'natural capital' and 'physical capital'. Well-meaning citizens from Global North and South alike believe techtransfer and digitalization essential for 'a just transition' to planetary sustainability. And indeed, the official response to the planet's metabolic breakdown is 'ecomodernist innovation' – a claim that engineered efficiency will de-materialize the amount of resources used by industry. However, automated production means more mineral extraction and heavy energy drawdowns, often undermining self-sufficient rural communities that already exist. The claimed reduction of material resourcing rarely factors in all operational aspects of the industrial production-consumption cycle. When fully researched, the ecological modernist expectation of progress simply does not hold up.

This is the ecological side of things, but social costs are locked in. For one thing, governments and international agencies treat decolonial and sex/gender politics as 'single issue' problems, thus disguising the intersectional complexity of power in a system of patriarchal-colonial-capitalist relations. In response to growing injustices, people targeted the World Trade Organization in a famous eruption known as the Battle of Seattle, 1999. This consolidated a 'movement of movements', which broad peoples' alliance held its first World Social Forum in 2001. In the decades that followed, a plethora of international networks would advance the grassroots conviction that 'Another Future is Possible!' On the streets of Davos outside the rich men's World Economic Forum and at UN Climate Summits, activists increasingly pursue sustainable political alternatives based on principles like subsidiarity and bioregional commoning. Academics have met in Barcelona, Leipzig and Budapest to discuss degrowth. The worldwide Rosa Luxemburg Stiftung works on Socio-Ecological Transformation. Yet in both Global North and South, among movement intellectuals and street fighters, there is a need for shared transversal consciousness-raising. It is time to hear from all the voices that have been marginalized and silenced as the Others of history. An ecological feminist analysis aims to restore that political integration by showing how workers, indigenous, womens and ecological politics are interlinked.

Androcene

Since the turn of the millennium, much public conversation has focused on the Anthropocene idea put forward by Nobel Prize laureate Paul Crutzen.[4] He argued that the invention of the steam engine and consequent Industrial Revolution destabilized the Earth's biogeochemical systems as a result of emissions from excessive fossil fuel use. A counter thesis from William Ruddiman claimed that 'settled agriculture' some 10,000 years earlier was what set in train the breakdown of planetary processes.[5] The late Will Steffen and colleagues would emphasize the global warming impact of technical innovation, consumerism and international trade as a 'Great Acceleration' occurring rapidly after the Second World War.⁶ Johann Rockström defined this state of CO_2-induced climate change, biodiversity extinction and a failing nitrogen cycle as critical transgressions of 'safe planetary boundaries'. He pointed out that more land-use changes and freshwater extraction will only further damage the Earth's systemic resilience. Rockstrom called for a return to biophysical thresholds existing in the Holocene era.⁷

Scholars in the Humanities and Social Sciences endorse the urgent need for scientific research of this kind but advise against 'naturalized' accounts of historical events such as the Anthropocene. As Noel Castree and colleagues explained, this label incorrectly implies that humans at large are responsible for the planetary crisis.⁸ Attribution of the Anthropocene to 'species activities' bypasses social factors like differences of sex/gender, ethnicity or class. In this way, the term 'Anthropocene' functions as an ideology, shielding those whose political actions and consumer choices are responsible for it. For example, some 100 multinational corporations are said to generate three quarters of the world's climate pollution, while Steffen et al. point to the concept of 'development' as a serious cause.

> The world's poorest countries, with a combined population of about 800 million people, have contributed less than 1 per cent of the cumulative CO_2 emissions since the beginning of the Industrial Revolution. However, the most recent data . . . show the dramatic changes over the past decade. For 2004, the emissions from developing countries had grown to over 40 per cent of the world total.⁹

And Steffen's figures are modest in the light of ongoing assessments. To sociological eyes, this trajectory stems from capitalist commodification of nature and the neocolonial spread of industrial lifestyles. Looking at modernity as a socio-economic system marked by an unequal exchange between Global North and South, Andreas Malm and Alf Hornborg put forward another label for the Anthropocene, namely the Capitalocene. Their formulation was further popularized in Jason Moore's account of the world-system.¹⁰ Given business as usual, Eric Swyngedouw and Henrik Ernston proffered the term Anthropo-obscene, to describe 'an immunological biopolitical fantasy that promises adaptive terraforming, an Earth system management of sorts that permits life as we know it to continue turning into a necropolitics for others'.[11] They named the California-based Breakthrough Institute as an exemplar of that hubris – a neoliberal think tank whose Ecomodernist Manifesto even entertains the possibility of 'a good Anthropocene'. Conversely, anarchist social ecologist John Clark places his analysis of the current era in the category of Necrocene.[12] Meanwhile, ecosocialist John Bellamy Foster argues that the scale of the Anthropocene Epoch is so vast that it should be subdivided into ages. Foster nominates the present as the Capitalinian Age, being an outcome of global monopoly capital post 1950. Nevertheless, he is hopeful that an ecological civilization may yet emerge:

> the necessary reversal of existing trends and the stabilization of the human relation to the earth in accord with a path of sustainable human development can only occur through social, economic, and ecological planning, grounded in a new system of social metabolic reproduction.[13]

The political literature now implodes with competing interpretations of the Anthropocene, with Kohei Saito's new volume *Marx in the Anthropocene* offering a comprehensive guide.[14]

Indeed the original Anthropocene label is problematic. If the scientists' idea of 'human species' error as cause of the global crisis is too easy, so is the Left's prioritization of 'capitalism'. The uncomfortable fact is that something is hidden, underlying each account. Marxists are right that the Eurocentric imaginary 'essentializes' humanity as a single powerful agent, overlooking the experience of other classes and races around the world. But the idea of 'humanity' is also sex/gender skewed – and by a margin of 50 per cent. On this basis, the present era might be more accurately described as an *Androcene*. Such a notion will reach beyond both natural science and political economy, into the very domestication of patriarchal-colonial-capitalist practices that have energized the global imperium for centuries. The *Androcene* refers to a worldwide system of social valuation that pivots on an unconscious competition between Self and Other. At its deepest level it embodies the *libidinal rift* observed in sex/gender politics whenever women's bodies are treated as *terra nullius*. But the wider historical outcome is an objectifying culture readily visible in ecocide, genocide, femicide and ongoing class war. Patterns of othering vary across cultures, but today the globally dominant form is the Eurocentric:

- Self   versus   Other
- Humanity   versus   Nature
- Masculine   versus   Feminine
- Production   versus   Reproduction
- Economy   versus   Ecology
- Capital   versus   Labour
- Rights   versus   Resources
- Mental   versus   Manual
- Clean   versus   Dirty
- Subject   versus   Object
- Mind   versus   Body
- Modern   versus   Primitive
- Order   versus   Chaos
- North   versus   South
- Land   versus   Water

The alternative to this objectifying dualist measure of value is a relational form of sociality grounded in identification rather than dissociation; as ecological feminists say: a sense of 'power with' rather than 'power over'.

The universities maintain the *1/0 imaginary* with the Two Cultures tradition, consigning Humanities and Natural Sciences to opposite sides of the campus. Recently, a form of posthumanist monism, influenced by Bruno Latour, would give the idea of historical agency to nature; but this object-oriented ontology is as apolitical as the Anthropocene story of the scientists.[15] The Humanity/Nature divide may be an old myth, but it is a contradiction that actively shapes peoples' identities and attitudes in everyday life. Even so, the dualism touches the consciousness of women, indigenous peoples, workers and youth, in different ways. An awareness of how this cultural conditioning works can help draw people together, rather than leaving them split into 'sociological intersections'. Ecofeminist attention to the material effects of cultural dualisms is a line of political strategy that can help unify the movement of movements.[16] Before taking this further, I want to share my grief for what is happening to Life-on-Earth – this passage comes from the close of my first book, *Ecofeminism as Politics*.

The Master Imaginary

In the beginning, the Judaeo-Christian tradition imagined nature as a Great Chain of Being. This was made up of a hierarchy of dominations from white men – and their God at the top; running down through white women; black men; black women; children; animal species; plants; air; water; and rocks. Capitalism and the Whig revolution gave middle-class men sovereign rights, but the liberal pluralist tradition got stuck after that. Socialism has so far failed to get working class men a fair piece of the pie. Feminism, post colonial struggles, ecology, in turn, each challenge the 'natural hierarchy' of privilege. But in this structure of patronisation, each link in the chain of dispossession squabbles with the others like anxious, unloved siblings. So Black and Green, or Red and Violet, sometimes lock in a battle for political recognition, arguing their case in the discourse of sovereignty and rights. The 'divide and rule' which follows, simply reinforces the master's control.

Now corporate colonisation through bio-prospecting literally undercuts the paternal deal and its domestic protections. For indigenous peoples and women, the semantics of self-determination is annulled by the thrust of white men's science beneath their very skins. Equal pay, contract law and land ownership walk a barren field, when blood, sweat and tears are mined and sold. Most Greens, ecofeminists, and Australian indigenous people, agree that the great culture of Europe is spiritually sick, economically and environmentally unsustainable. Wilderness in measured doses – usually on Sundays – has been a salve to the West's malaise, but like sexuality and conflict, it must be contained. The flaccid, unfocused faces of the brotherhood in suits tell it all. If the wild does indeed reveal the crisis of a bankrupt civilisation, then let's keep working away at that point of least resistance. Green, Black, and ecofeminist politics will each benefit by supporting men's own movement struggle to leave behind the ugly capitalist patriarchal default position.

We do not challenge the monoculture of corporate savages by denying 'difference' and the 'wild'. To dub talk of indigenous specificity as 'primitivising' is to take on board the nineteenth century evolutionist mentality with its ladder of 'progress', rather than to respect the diversity of peoples' ways. It is to shun 'women's business' in favour of 'white men's business'. Just because an indigenous people has no division between Culture and Nature, does not mean they are closer to Nature in the Western sense of inferior. This is also true of women under capitalist patriarchal ideology. Indigenous cultures offer rational models of how that linkage has been realised by peoples in their history. To say this, is neither romantic nostalgia, nor a perverse claim that indigenous environmental practices are 'better' than 'modern' scientific ones. All cultures make mistakes. Neither is it to suggest that indigenous peoples should necessarily want to live as their ancestors have – though some may and some whites along with them. All cultures grow and change.

In times such as these, when the lifestyle of a few brings ecological devastation and poverty to many, there is an urgent need to reappraise economic alternatives to industrialisation; to re-frame our political tenets; and to start taking small every day steps away from the folklore of self loathing and its pitiful gadgetry. Eurocentric concepts, like management and control, paternalistic beliefs that Western technologies are essential to the good life, these are the most insidious forms of invasion. After all, what is this modern civilisation but a time and energy consuming process of dismantling living things and turning them into dead matter? The only thing that increases with 'economic growth' is waste. And so a holocaust goes on among us: dammed up rivers run sour and parched soils crack open; continents swarm with environmental refugees; man-made viruses are unleashed; silently, an ozone hole and electromagnetic radiation cull new cancer victims; oil spills suffocate sea life and melting ice plateaux threaten islander peoples. Will you too close your eyes to these crimes, the linear model of development exported by an Enlightened West?[17]

That was written twenty-five years ago now, but postmodern predictions of 'the end of history' and its grand narratives have not been realized. The hubris of mastery is stronger than ever, armed now with the vanities of digitalization and AI. Creative labels for the new era continue to proliferate, although Franciszek Chwalczyk's survey of the Anthropocene literature shows that scarcely any studies of the construct apply a sex/gendered sociology of knowledge in their analysis.[18] Many women have a trenchant critique of the *Androcene*, grounded as their labours are in the protection of natural living cycles. An ecofeminist politics moves beyond the dissociated cultural boundaries of sex/gender, ethnicity and class. Emerging across several continents some five decades ago, ecofeminist political interventions were womanist, decolonial and socialist from the start.

Patriarchal-colonial-capitalism

The contemporary Left usually talks as if capitalism is *sui generis* and a self-reproducing system. The argument to be made in this book is that the dialectic of patriarchal-colonial-capitalist practices forms a single whole, with each subsystem internally related to the ones that preceded it historically. Masculinist domination, extending back thousands of years, is the first level of this *Androcene*, setting humans over and above the natural world often in the name of religion.[19] It seems that atleast in the globally dominant culture, 'In the beginning was the mother' is abstracted from its materiality to become 'In the beginning was The Father'. This ancient sex/gendered dissociation, a *libidinal rift*, effectively created 'women' as the second sex and first colony. The next level of *Androcene* othering would extend outwards across nature's territories with the appropriation of lands and enslavement of bodies. The history of colonization extends from nomadic tribal wars, to biblical Exodus, to Ancient Rome, to the Atlantic Triangle and on to the twentieth-century United States. *Androcene* capitalism is only 500 years old – a dual class system but still making good use of the earlier systems of domination. As civilizations become sedimented over centuries of social interaction, norms are habituated, internalized and recharged daily with embodied energies. Institutions such as the nation state, capitalism or science emerge in this collective sublimation as a 'second nature'.

Neither liberal democracy nor socialist revolution can release the mythic order of Man over Woman, White over Black, Humanity over Nature. Materially, the patriarchal-colonial-capitalist complex uses these objectified dualisms to enfold, indeed lock, one subsystem into the next. In a reflection on the role of dualisms, the Argentine decolonial theorist Walter Mignolo once wrote that 'Linear global thinking is the story of how Europe mapped the world for its own benefit and left a fiction that became an ontology: a division of the world into East and West, South and North, or First, Second and Third'.[20] Until recently, much postcolonial theory was poststructuralist and assumed discourse to be the key determinant of social life. In fact, the Latin American Subaltern Studies Collective was said to ride with 'the four horses of the apocalypse' – Foucault, Derrida, Gramsci and Guha. But scholars like Maria Lugones and Annibal Quijano would reappraise the relevance of European thinkers.[21] With time, some decolonial thinking has merged with the global wave of an ecofeminism that speaks the experience of everyday women.[22] This shift has helped grow alliances between colonized farmers in the Global South and activist housewives in the Global North. Afterall, each comes to politics from a zone of non-being – a *terra nullius*. Metaphors like 'mining virgin lands' or 'the rapacity of capital' are not coincidental; they are symptomatic of the *Androcene* unconscious.

Across the world, among elites and movement cadres alike, sex/gender consciousness-raising is long overdue. Anthropologist Arturo Escobar from Colombia reaches into this layered politics and speculates on its consequences for Life-on-Earth.

> At one end we find matristic, convivial, futuring and, broadly speaking, relational visions that highlight the re/creation of worlds based on the horizontal relation with all forms of life, respecting human embeddedness in the natural world. At the other end of the spectrum there lies the dream, held by the flashy techno-Fathers of the moment, of a post-human world wholly created by 'Man'. This is the world, for instance, of synthetic biology, with its gene-centric view of life; of booming techno-alchemies for genetic

enhancement and the prolongation of life; of robotics, cyborgian fantasies, space travel, nanotechnology, unlimited 3-D printing, and much more; of the bizarre geo-engineering schemes concocted in corporate boardrooms as solutions to climate change; and of those advocating for the 'Great Singularity', a technologically-induced transformation 'when humans transcend biology', where life would finally be perfected, perhaps as in the world-without-mothers of AI fictions such as those portrayed in the film *Ex-Machina*, where women's ability to give life would finally be completely usurped since wo/man would be wholly created by men through the machine.[23]

Escobar is describing the inevitable outcome of a patriarchal culture, whose negation of the mother-body is its first step. For this reason, it is exciting to encounter Leonardo Figueroa-Helland, another decolonial theorist, describing his path to a regenerative politics as 'a rematriation'.[24]

Global Debt Matrix

The *Androcene* narrative, premised as it is on an ancient Humanity/Nature rift, continues to underlie what passes for common sense.[25] Emerging from the world's great monotheistic religions and reinforced by classical Greece, the Chain of Being hierarchy has been elaborated over centuries to justify the downward force of power. From Gods and kings, on to presidents, entrepreneurs, technocrats, managers, workers and husbands. With the European Enlightenment, the theological narrative became lodged in the plausibility structures of modern economics, law and science. Today, most people still take for granted the contradictory alienations of the master imaginary. Even in the modern so-called 'developed' world, this preconscious *episteme* frames the ego, organizing experience as 'a reality' in two parts – valued and not valued; a reified 1/0 grid is spread across the world. Negatively attributed, abstract Nature and its human synonyms are merely 'reproductive', while Humanity proper accords himself the right to make 'productive' use of slaves, workers, peasant and indigenous peoples, mothers, youth, animals, plants, air, water and soil. The primitive Promethean capture of fire, aka energy, would still appear to be the ultimate grail, so no surprise that the planet is burning.

Reading Markus Kroger's travels into the Brazilian Amazon, one of his most shocking encounters is with a millionaire landholder occupying part of the decimated forest territory of Mato Grosso. Typically, this agro-industrial farmer cultivates acre upon acre of land cleared and flattened as far as the eye can see. What grows there are monocultures in rotation each year – usually Bayer GMO soy, cotton and corn. Neighbouring farms raise genetically modified pigs, cloned for rapid maturation; egg producers dispose of market excess male chicks by beheading, electrocution or simply suffocation in plastic garbage bags for landfill. Such is the dissociated design of global capitalism and its monocultures. Water and forests are sold off as commodities, and mineral extraction across South America has grown by 1,000 per cent in recent years. But as Kroger writes: 'once the "resources" have run out, what remains is machineries, waste, and barren landscapes.'[26]

Under the patriarchal-colonial-capitalist system, all life-giving energies circulating through nature are appropriated upwards through the *Androcene* hierarchy.[27] The outcome is a global matrix of thermodynamic debt. The categories of this extractivism are multiple.

- Workers and slaves, men or women, white or Black, have a 'social debt' owed to them by the capitalist employer's systematic underpayment of their labour time.
- Indigenous foragers, peasant smallholders and displaced refugees have a 'livelihood debt' owed to them for appropriation of their lands and skills, under the modern form of colonization known as 'development'.[28]
- Women have an 'embodied debt' owed to them for daily services freely provided like domestic care, sexual gratification and/or reproduction and even surrogate pregnancy.
- Youth have a 'generational debt' owed to them by a global economic system that transfers the collateral costs of the patriarchal-colonial-capitalist imperium into the future. The sexualized predation of children is another generational theft.

- A 'species debt' occurs as animal and plant lives are taken for food, clothing, transport, manufacture and recreational hunting. Meanwhile, biodiversity is re-engineered into commodities from medical products to biological weapons.[29]

- The entire system of life-regenerating flows between Outer Space and the Earth is disabled by *Androcene* indifference to nature's metabolism, setting in motion entropic disintegration as a full-scale ecological, if not, 'planetary debt'.

As outlined in the ecofeminist anthology *Eco-Sufficiency and Global Justice*, this global debt matrix is at once moral and economic, thermodynamic and genetic, and it provides a powerful rationale for a transversal movement politics.[30]

As things stand, each material debt drives a particular social movement, and this divide and rule is convenient to those in power. Neoliberal governments easily sideline sectoral protests over exploitation as competing forms of 'identity politics'. In the online media, expressions of righteous outrage are dismissed as 'political correctness' or, worse, 'conspiracy thinking'. True, violences experienced across the matrix of appropriation are unique, but they are also shared as part of a long-evolved world system of unequal exchange. Many insightful youth activists already connect the dots here, for it is plain that the twenty-first-century theft of life energies is a lethal system.[31] However, if atmospheric-biospheric breakdown results from an instrumental 1/0 logic, the dominant ecomodernist approach to remediation operates by the very same model.

Transversal Praxis

In the twentieth century, the study of political economy gave way to political ecology as more and more social problems triggered environmental damage.[32] Political ecology is transdisciplinary, often blending insights from psychology, cultural studies, anthropology, sociology, economics, geography and biology. Political ecology calls for thinking down through layers of taken-for-granted sociocultural sediment. Indeed, when patriarchal-colonial-capitalist effects are found to be mutually reinforcing, the explanation needs a model of 'overdetermination'. While standard ecomodernist analysis is linear, instrumental and positivist, political ecologists will adopt theoretical vocabularies from psychoanalysis and neo-Marxism, as well as concepts like South Africa's *ubuntu* and Ecuador's *buen vivir*.[33] The essay collection *Pluriverse: A Post-Development Dictionary* reveals further 'ways of worlding' that are fair and sustainable.[34] For decades, peoples have been striving to protect their cultures against pressures to universalize. And reinforcing this move are many committed intellectuals and activist networks like Global University for Sustainability, the Environmental Justice Atlas and Radical Ecological Democracy.[35]

Across class, ethnicity, sex/gender and age, people begin to be aware of their position in the world system and the personal advantage, or otherwise, that comes with it. This reflexivity is essential for newly developing deliberative approaches to democracy.[36] As peoples worldwide join up to organize life-affirming societies, they gain a clear sense of how their differences are simply historically created. Women's reproductive biology is an exception to this social liberation however, because it must speak of both biophysical 'sex' and culturally constructed 'gender'. But growing a liveable planet from ground up means learning to think both/and, stepping away from old dualist assumptions like Body versus Mind shaped by the Eurocentric master ego. A sense of self as materially continuous with the natural world is basic to this renewal, and hybrid political standpoints, such as an ecofeminism, postcoloniality, ecosocialism or transgender, already signify this rising consciousness. Both the geographic and domestic margins of the patriarchal-colonial-capitalist complex prefigure new kinds of reciprocity in the humanity-nature nexus and new understandings of matter-energy exchange.[37] In the final analysis, it is time to concede that all 'humans are nature-in-embodied-form', and if there is to be a future for Life-on-Earth, it will depend on recovery of that relational awareness by the coming generation.

As the *Androcene* spreads extinction across the world, movements for social change are in constant conversation and often competition with each other. Understanding this radical politics means tracking in and out of their literatures and debates. It means revisiting eco-Marxists, decolonial thinkers, deep ecologists or posthuman feminists. This book, part of a trilogy, does that – but always looking to the future of Life-on-Earth. The *Androcene* – the patriarchal-colonial-capitalist system as we know it today– evolved over time as a unity, and these three books spell out key aspects of its entangled practices. Meanwhile, the Others of *Androcene* culture – women, indigenous peoples, workers and the ecology movement – all struggle with different facets of this force, so a shared understanding is vital to their eventual political success.

- *DeColonize Ecomodernism!* starts with the wordwide *Androcene* and finds the old colonial ideology of *terra nullius* still active today as government policy, and international agencies tear apart indigenous peoples' life-worlds and their time-tested capacities for protecting nature. The chapters spell out the biopolitical violence of digitalization and genetic engineering and trace two decades of creative defiance by global peoples' movements against the contradictions of ecomodernist development.

- *EnGender EcoSocialism!* suggests that the now global *Androcene* culture was set in place by an originary *libidinal rift*. Thus an emerging ecosocialist theory needs to look deeper than the 'metabolic rift' that capital causes between humanity and nature. An integrated ecosocialism needs an 'embodied materialism' to heal the *libidinal rift* in its own theory and praxis. Denial of this level of politics will hold back a rounded theory of labour that serves workers, indigenous peoples, women and nature alike.

- *ReGround Ecofeminism!* reveals how some feminist theory, influenced by the androcentric *1/0 imaginary* and its dualisms, undermines women's efforts to voice their everyday experience and activist efforts to protect nature. This academic 'ungrounding' can destroy the postcolonial unity between women of Global North and South as well as their ecosocialist standing as a 'reproductive labour' class.

The trilogy draws on my time as a scholar-activist involved in radical politics locally and globally over several decades. Much of the material is new, some of it revisits themes from previously published work, but each volume is written to appeal to the imagination of new activists. After all, those who do not learn from history are likely to repeat it!

Speaking in 2021 at a Youth4Climate conference in Milan, eighteen-year-old Greta Thunberg mocked world leaders over their years of fake climate promises and tired slogans; it has been 'just 'blah, blah, blah' and more 'blah blah blah'. And she went on:'The world's planned fossil fuel production by the year 2030 accounts for more than twice the amount than would be consistent with the 1.5C target. This is science's way of telling us that we can no longer reach our targets without a system change.'[38] With the world 'speeding in the wrong direction' young protesters from every nation are confronting policymakers to examine alternative solutions. A political strategy that encompasses sex/gender, ethnicity and class simultaneously may test older activists used to working on single-issue agendas. But facing down the patriarchal-colonial-capitalist complex calls for fresh energies. This political synthesis is critical to resolving the plight of a generation whose future looms dark beyond imagination. As their protest posters have it: 'You know its time for change, when kids act like leaders, and leaders act like kids.'

2

Global Synergies
Livelihoods or Lifestyles?

If the first-generation Green thinkers like Rudolf Bahro and Wolfgang Sachs brought a new awareness to European activists, political writing from the global peripheries had the edge on them.[1] Mahatma Gandhi's advocacy of economic sufficiency is honoured today by India's Vikalp Sangam *swaraj* communities.[2] From Mexico, renegade priest Ivan Illich exposed the ecological incoherence of industrial societies in *Energy and Equity* and offered a way out in *Tools for Conviviality*. Gustavo Esteva would carry these insights forward alongside the Zapatista liberation movement. At the first World Social Forum in 2001, Pablo Solon advanced the Andean economic model of *buen vivir* as a Systemic Alternative.[3] In the United States, the idea of bioregionalism was already popular among environmentalists thanks to writer Kirk Sale and activist Peter Berg.[4] Designs for eco-sufficiency through permaculture farming originated with Bill Mollison in Tasmania – an approach echoed in Ted Trainer's tract *Abandon Affluence!*[5] In South Africa, according to Mogobe Ramose, the Zulu, Xhosa and Ndebele peoples say: *Umuntu ngumuntu ngabanye Bantu* when talking about humans as part of the biophysical web. There is no dissociation in the *ubuntu* psychology: 'We are, therefore I am . . . by harming others, I harm myself.'[6] In related vein, indigenous American Potawatomi activist Kyle Whyte has called for 'self-limits'.[7] Columban missionary Sean McDonagh converted to a new ethic while living among the T'boli highlanders of

Mindanao; theologian Leonardo Boff's *Cry of the Earth* celebrated the wisdom of indigenous economies. Christian decoloniality would culminate with the Papal encyclical *Laudato Si': On Care for Our Common Home* (2015).[8]

Yet few visionaries, secular or religious, have done justice to the foundational question of power in 'the fatherworld' – and its seminal role in the evolution of patriarchal-colonial-capitalist institutions. The political versus personal divide marks other important agendas too, from EcoSocialism and Social Ecology on the Left to Deep Ecology on the Right; so women have been framing an alternative political ecology. Dianne Rocheleau's case studies of women's sustainable practices around the world would break through that old blind spot.[9] The rewriting of political ecology by ecofeminists is a further attempt to remedy it. The late Maria Mies' classic thesis in *Patriarchy and Accumulation on a World Scale* was one of the first to foster a synthetic grasp of what I call the *Androcene*. Inspired by Rosa Luxemburg's profound analysis of imperialism and by close observation of rural women's work, Mies together with Vandana Shiva catalysed a new synthesis of feminism, decolonialism and socialism. Their materialist focus is the economics of regenerative labours North and South and a critique of modernity as 'mal-development'. The ecofeminist approach to 'livelihood' is known as 'the subsistence perspective' emphasizing that an economy should be about what people need, not about other things.[10]

The Activist

If the spinning wheel was the symbol of Gandhi's cultural resistance to British rule, so, in an era of corporate biocolonization, 'the seed' is ecofeminist Vandana Shiva's symbol of an Earth Democracy. And just as Gandhi's politics was about economic self-reliance and cultural integrity, so the focus of Shiva's ecofeminism is the affirmation of life in all its forms. As she observed in the early days of her ecofeminist journey, the protection of biodiversity and protection of cultural diversity go hand in hand. But with neoliberal globalization, a monoculture takes over – an enclosure and commodification of living resources that knows no bounds. First, forests are enclosed, then farm lands, then rivers and then the very flesh of seed itself is turned into saleable property in a ruthless competition for profit driven by a few powerful men. Shiva dates the beginning of biocolonization from 1995 with the World Trade Organization's (WTO's) Trade Related Intellectual Property Rights (TRIPS) legislation. The patriarchal-colonial-capitalist solipsism of TRIPS legalized the corporate biopiracy of indigenous intellectual property simply by terming it Trade Related (Article 27.3b).

After years of writing and resistance on behalf of livelihood, Shiva established *Navdanya*, a learning centre at Dehradun. *Navdanya*, meaning Nine Seeds, is also the name of an India-wide network of seed-saving farmers, mostly women. She has helped build the worldwide movement of movements against globalization, working with local peasant groups and transnational ecofeminists like Diverse Women for Diversity. At the same time, she has written innumerable books since her first ecofeminist foray in *Staying Alive* (1989) – most recently, *Oneness vs the 1%: Shattering Illusions, Seeding Freedom* (2018).[11] What follows is a discussion of *Earth Democracy* (2006) – a synthesis of reflective fragments and impassioned addresses given by Shiva over many years. What is most vibrant and fresh are her accounts of global political activism. For example, with Linda Bullard of the International Federation of Organic Agriculture and Magda Aelvoet of the European Greens, Shiva successfully pursued a ten-year legal battle at the European Patent Office against R. W. Grace and the US Department of Agriculture, which had patented the *neem* tree, a traditional Indian pesticide, as their own invention. Another campaign reclaimed the germ line of basmati rice from patents held by the Texas company RiceTec. In 2004, Shiva's Research Centre for Science, Technology and Ecology, working in tandem with Greenpeace, managed to roll back a potential 'terminator technology' – packaged infertile planned obsolescent seed in Monsanto's patent of a soft milling Indian wheat. Each of these political interventions liberated life forms and local knowledge from the dead weight of corporate food fascism.

The movement struggles at Seattle and later Cancun, and victories of the common people 'for life before commodities', are well known through media reports; but many ongoing forms of resistance to the capitalist monoculture are less well known. For example, world food supplies may be colonized and emptied out by the likes of Cargill, Monsanto, Phillip Morris, Nestle, but the subsistence alternative and 'slow food' movement flourishes under the name of *Terra Madre*. Shiva describes its vibrant 2004 gathering at Turin, Italy:

> there was a community of dried mango producers; entomophagous women of Ouagadougou, Burkina Faso – women who harvest and sell edible insects; the Boabab community of Atacora, Benin; basil growers; makers of Liguria; nomadic shepherds from India and Kirgbity; sheep breeders from Central Asia; jasmine rice producers from Thailand.[12]

She urges us to grow *Earth Democracy* by keeping the pressure on global institutions and markets with *bija satyagraha* – civil disobedience against patenting seeds. The results of such public vigilance can be seen in 2004 Swiss laws requiring corporate peddlers of genetically modified product to observe the precautionary principle and to be liable for human and environmental damage under the polluter pays principle.

Shiva's argument moves from the theft of seed to water enclosure for privatization and sale – hydropiracy. Like air and soil, water is a free good, and access to these commons is a fundamental human right. These are surely matters of 'sovereignty' being undermined by the WTO's neoliberal free trade regime. But India is beset by World Bank-orchestrated, government-compliant water projects. Consider the Tehri Dam at the source of the Ganga: its construction has been declared unsafe by the International Commission on Large Dams, and its failure would destroy cities, farmlands and lives across the Gangetic plain. As if this were not enough, an emergent Citizens Front for Water Democracy grapples with a further technological pipe dream – the engineered rerouting and joining of innumerable rivers in the Himalaya hills and others criss-crossing Peninsular India. This exercise is a prelude to corporate privatization of water supply, but the devastation of local ecological habitats and human communities by such irrational capital works is an alarming prospect. The Indian term for water democracy is *Jal Swaraj*, and again, Shiva tells how peasants and women lead in protection of it. In *Plachimada* hamlet, Kerala women protested against loss of their water supply to a local Coca-Cola factory, and their case for livelihood was affirmed by the court. Their success has now set a precedent for eighty-seven more grassroots movements against Coca-Cola and Pepsi. The critical question is:

How do we achieve global political recognition for these new material forms of sovereignty?

Shiva notes that economic globalization and cultural nationalism are two sides of the one coin. Thus,

> When economic dictatorship is grafted on to representative, electoral democracy, a toxic growth of religious fundamentalism and right-wing extremism is the result. This, corporate globalization leads not just to the death of democracy, but to the democracy of death, in which exclusion, hate, and fear become the political means to mobilize votes and power.[13]

Ecofeminists see the convergence of market fundamentalism and religious fundamentalism resulting in an overwhelmingly masculinist global value system. It will be a race against time to keep this hegemonic force at bay as the movement of movements and everyday folk, North and South, struggle to build rational life-affirming global institutions. Shiva believes that 'the commons are the highest expression of economic democracy', and she calls for a Water Parliament – or federation of same.[14] She does not use the term dual powering, but her vision is anarchist and populist. And if she would reclaim the nation state as a necessary lever in building an *Earth Democracy*, nevertheless, sovereign control must remain at the level of local communities.

The Teacher

Vandana Shiva's book finds its complement in the Global North with Patrick Hossay's *Unsustainable: A Primer for Global Environmental and Social Justice* (2006).[15] Hossay teaches at the Richard Stockton College of New Jersey and is engaged in development projects in Central America. He challenges those who oversimplify 'The trouble we're in', arguing that global population size is not the problem, for example, but food distribution is. To support this, he cites the American Association for the Advancement of Science to the effect that 78 per cent of all malnourished children live in countries with food surpluses.[16] His primer is full of compelling graphs and statistics; the first chapter is a veritable *tour de force* on the question of climate change. It is worth reading

Hossay for this thoroughly documented argument alone. We have sky-borne carbon, methane, HFCs; unwelcome emissions from peatbogs, cars and cattle; hurricane events, melting ice and flooded cities; desertification and landslides; destabilized habitat; new predator chains and disease viruses; displacement of human communities and starvation.

Like Shiva, Hossay is deeply concerned with the despoliation of biodiversity and of water, and he notes the implication of Big Pharma in both. The US drinking water cycle is routinely contaminated by antibiotics, Prozac, pain relievers and hormones from birth control pills. Likewise, the use of water as an industrial sewer brings toxic heavy metals to streams and aquifers. On the general wastefulness of high-tech provisioning, Hossay cites Shiva: 'making a single six-inch silicon wafer requires 2,275 gallons of water; so, a normal-sized computer manufacturing plant requires 236,600,000 gallons of water each year. Efforts to expand high-technology industries in India have dire potential for the country's water crisis.'[17] But when Hossay says 'our' and 'we' he means North American readers, and this taken for grantedness can be disconcerting for those in the often invisible Global South – not to mention inhabitants of the economic-South-in-the-global-North as so many women are. Nevertheless, his target audience of Middle America is appropriate in the sense that the onus is on these energy-greedy high-tech consumers to wake up to their superordinate global footprint. Hossay intimates something of this class, race and sex/gender complexity in connection with waste disposal and environmental racism; but on the whole, his text is not strong when it comes to sociological reflexivity. For instance, despite gains won by African American communities through the environmental justice movement, massive amounts of the US waste are still exported to Third World states. This form of class and race abuse gets past the Basel Convention by labelling the waste recyclable.

Hossay asks: 'How did we get in this mess?' and here he gives a non-sex/gendered account of Sir Francis Bacon and the rising ideology of Enlightenment science; the barbarism of nineteenth-century colonization by European states; and the unprecedented international influence of the United States today. His treatment of the origin of the post-war 1945 Bretton-Woods financial institutions – IMF, World Bank and now WTO – is clear and concise. He goes on to explain how Nixon's release of the US dollar from the gold standard set a cumulative economic force in motion and how this was followed by an unhinged speculative cyber-economy. Hossay's anecdotes on the manipulation of global market policy by successive Washington presidents are illuminating, as is the League of Conservation Voters' list of early administrative appointees by Bush Jnr.

- Chief of Staff = Former President of the Automobile Manufacturers Association and Chief Lobbyist for General Motors
- National Security Advisor = Former Chevron Board Member
- Chair of the Council on Environmental Quality = Lawyer for General Electric
- Secretary of the Interior = National Chair of the Industry Group representing timber and chemical manufacturers
- EPA Deputy Administrator = Former Vice President of Government Affairs at Monsanto.

This makes it very plain why the United States stalled on the Biodiversity Protocol and Kyoto and why free compacts like the North American Free Trade Agreement (NAFTA) have such eco-unfriendly content.

The impressive collection of facts and figures in Hossay's *Unsustainable* backs up Shiva's observations on rural impoverishment. A Worldwatch graph shows farmers' share of food profits declined from near 40 per cent in 1910 to less than 10 per cent in 2000. The profit share of food marketing companies increased from just over 40 per cent to nearly 90 per cent in 2000. No surprise then to find thousands of farmers leaving the land each year or to learn that many commit suicide, unable to meet loans for agro-tech seed and equipment. Hossay details the World Bank involvement in rolling forward innumerable destructive projects, and he outlines the related role played by Export Credit Agencies. One such corporate assistance programme, the planned settlement of Javanese, ostensibly as labour, in resource-rich Papua now exposes its full ecocidal and genocidal cost, as boat loads of brutalized refugees try to make it across the straits to Australia's northern coast. Hossay defines modernization as 'adoption of the social, political, and economic systems of the wealthy, as well as the acceptance of their cultural values and priorities'. He goes on to observe that in practice, modernization has meant that slavery is more common now than even at the height of Transatlantic slave trade.

When it comes to addressing what is to be done, some may be disappointed at the modesty of these proposals. Caring people from opposite sides of the Earth can work together to make change. Yet, environmentalists from affluent nations are not always aware of the depth of repression enacted upon First Nation peoples and ways of worlding.[18] While Hossay is adamant that individual lifestyle changes like recycling and car pooling are not enough, his big-picture appears to be a sort of environmental welfare state – whereby public institutions exert control over unruly corporations. Given the ubiquitous revolving door of capitalist political economy, this idealistic welfarism on the part of global elites is not a convincing scenario. When Hossay writes about half the Earth's population surviving on less than $2 per day, he appears to adopt the monocultural logic of the UN, IMF and associated financial institutions. Shiva's *Earth Democracy*, written from the Global South, prefigures a stronger materialist alternative. Her decolonial ecofeminist approach to protection of biodiversity and cultural diversity prefigures an Other way. By this logic of the commons, poverty is revealed to be not so much about getting money, commodities or even jobs in the waged labour sense. There will never be enough of these things to go around under capitalism, because the system itself is premised on the creation of scarcity. In an *Earth Democracy*, what counts is the human right to livelihood – sovereign access to land, water and the free seed of nature's bounty. This ecological reckoning calls for a fundamental epistemological break from the patriarchal-colonial-capitalist monoculture as we have known it.

A later Vandana Shiva book – *Oneness vs the 1%* (2018) – updates Hossay's list of the US ruling-class figures, with a focus on the philanthro-capitalism of the Bill and Melinda Gates Foundation. It turns out that philanthropy can serve as a tax dodge. More significantly, Shiva points to the foundation's role in the war against cash along with Microsoft, E-Bay, Mastercard and Visa. Just as Rockefeller pushed the Green Revolution into oil-based petro-farming, so the Digital Revolution will be based on bytes. Shiva reports that the Indian Ministry of Finance has already established a Cashless Payment Partnership, and British and Australian prime ministers were each personally lobbied in a brief personal fly-in/fly-out visit from Gates. The question is: How safe will life savings be under a privately owned and electronically precarious global 'securitization' system? A further concern is the revolving door between former Gates Foundation employees and Big Pharma such as Pfizer, Genentech, GlaxoSmithKline, Merck, Novartis and Monsanto. The interaction with Big Fossil, such as Anglo American, BHP Billiton, Glencore, Shell and Chevron, is similarly concerning. The political and technological convergence of an energy-hungry information industry with a nature-hungry genetics industry tightens the stranglehold of ecomodernism over everyday life.[19]

3

Terra Nullius

Consuming Lands and Bodies

The Earth Summit at Rio de Janeiro in 1992 set the stage for political struggles now unfolding in the twenty-first century. It formalized the influence of transnational corporations within the United Nations and other multilateral agencies. It set up new global governance measures through the Climate Change and Biodiversity Conventions, and it laid down guidelines for extractivism. The new wave of globalization would be based on corporate resourcing of minerals and biodiversity for engineering into new products and services. The process was formalized in four steps – Resource Assessment, Regional Agreements, Conflict Resolution and Intellectual Property Rights. The Rio *Agenda 21* was actively sponsored by the Business Council for Sustainable Development with outcomes to be advanced through free trade provisions of the World Trade Organization (WTO).[1] Regional associations would help transnationals penetrate resource-rich localities. In Australia, a new Municipal Conservation Association tuned into Ecologically Sustainable Development (ESD) principles, readily lent itself to this. At the World Bank and the International Monetary Fund the language of 'development aid' would be replaced by 'economic cooperation'; or if stronger inducement was required to involve the Global South, then Structural Adjustment was the word. At the same time, an ideology of Green Business was spreading through right-wing think tanks, environmental front groups and appropriate cultural activities.

Thus, Hydro Quebec, which had forced thousands of Canadian First Nations from the Hudson Bay area, set up a university Chair of Environmental Ethics. In Australia's Northern Territory, the Ranger Uranium mine funded Aboriginal Studies at the university in Darwin. This move that was very divisive of indigenous political loyalties, given the company's links to France's racist nuclear testing programme in the South Pacific.[2]

Extractivism

An intensified capitalization of indigenous struggles would take place very rapidly following the Rio Earth Summit. In Central Australia, at Alice Springs, a federal government Landcare program and the Indigenous Central Land Council hold a digital catalogue of mineral, biological and cultural resources. Local leaders are assured that resource assessment based on geographic information systems (GISs) will enhance Aboriginal livelihoods by enabling resource management in such a way that genetic and mineral items can be extracted while 'ecological balance' is maintained. Tribal Elders have been persuaded to lead resource assessment researchers into special sites, to be mapped after extensive consultation and cross-checking. But as formerly 'unchartered sites' are translated into the Eurocentric techno-ontology, the original colonial assumption of empty land – *terra nullius* – is turned on its head. The mapping process is said to 'marry' two information bases, wherein 'cultural data' is supposed to remain with local people. The claim is questionable, as are many aspects of this extractivist programme from the perspective of cultural autonomy and survival. If Landcare requires open access to all data gathered, and if the same information can be represented either in 'bush tucker' or in scientific terms, there is, in fact, no way of protecting indigenous intellectual property.

The debate over environmental racism is beginning to focus on problems like these. But some people expected resource assessment to help 'identify' regions suitable for politico-legal negotiation between government, corporate and indigenous interests. Could Regional Agreements be the way towards sovereignty? Few have asked, what role transnational firms might have in steering this 'regionalism'? Political economist Greg Crough working with the Northern Australia Research Unit in the 1990s was frank: basically a resource assessment gets done, then native title on inalienable freehold land is traded away for royalties. This view was contested by Les Carpenter, a well-known negotiator of the Inuvialuit Final Agreement in Canada and ambassador for Indigenous Regional Corporations. Similarly, Daryl Pearce, Northern Land Council director, was advising Aboriginal communities to 'get hold of a lawyer and negotiate a deal'. Go straight to the bottom line, according to Pearce, 'make use of contract law, it's purely about business'. But can indigenous peoples negotiate fair deals with mining companies that are 40 per cent offshore owned? Should self-determination hang on economic growth? Some Aboriginal pastoralists and tourism entrepreneurs operating under license from the Northern Land Council have been optimistic.

Against the profusion of legal remedies applicable to First Nation freedoms in Australia, Regional Agreements have seemed attractive to many. Others are not so sure. At the 1998 Ecopolitics IX gathering in Darwin, Michael Mansell of the Tasmania-based Aboriginal Provisional Government pointed to the fact that the Native Title Act 1993 covers only 3 per cent of First Nation peoples on the continent. Mick Dodson, who was Aboriginal and Torres Straits Islander Social Justice Commissioner at the time, observed that Regional Agreements might even undermine Mabo, the historic High Court decision on indigenous land ownership. Warning his people not to give up Native Title rights to get basic 'citizenship rights' like health and education, Dodson urged: 'keep on about human rights so as to access the force of international conventions' signed by the federal government. Of course, the very concept of 'rights' is corrupted by its origin in the individualistic and adversarial ideology of bourgeois rationalism. By contrast, indigenous ethics are communitarian, emphasizing mutual support and exchange rather than possession. Moreover, the very idea of one person 'granting equal rights' to another person is simply an oxymoron, and it typifies the delusional aspects of the liberal political tradition. The very act of 'giving rights' underscores the colonial moment of loss, so taking away as much autonomy as it bestows.

As the Rio *Agenda 21* created opportunities for indigenous elites, it was weakening community solidarity. At the same time, the approach would bring generous career opportunities for middle-class white consultants, including one or two liberal feminists. International treaties and conventions need data collection, analysis and reporting; legislation; and compliance monitoring. Healthiest growth industry of all is 'dispute resolution', the latest tool of government and industry for containing and managing conflict while creating an appearance of reasonableness and open consultation. These psychological techniques, applied to key tribal 'players', include mediation and assisted negotiation, but they are of little use to people when the sociological power and privilege of 'partners in dialogue' is not made explicit. The corporate lingo of 'harmonization' and 'partnership' thinly veils greed and patronization. As Martie Sibasado from the Kimberleys summed up the frustration of one such mediation session: 'Why do you have to leave at 3 o'clock to catch a plane, when I've had to walk all my life?'[3] So-called capacity building – assessment, monitoring, management and dispute resolution techniques – stud the colonial discourse of the Business Roundtable and UN agencies. Capacity building, also called 'enhancement', is the export of 'universal' – read white masculinist – skills needed to ease 'technology transfer'. There must be a better path to sovereignty than trading Black cultural meanings for middle-class white men's rights.

Exterminism

Indifference to the rights of First Nation people in Australia is ongoing and takes many forms: from deaths in custody of young Aboriginal men, to the community intervention programme which sent the military into the Outback to sort domestic issues, to the planned nuclear waste dump at Kimba near people's homelands, to the bombing of Juukan Gorge 46,000-year-old heritage site by Rio Tinto mines.[4] Perhaps an encouraging sign that emerged during the debate over Regional Agreements was a new form of grassroots localism brokered by women of the Melanesian Environment Network and the Australian Conservation Foundation.[5] However, the intersectionality of sex/gender, race, class, age and species politics is complex. To take 'the woman question': just as indigenous peoples captured by coloniality are made to believe they cannot live adequately without 'modernity', so women even in the 'developed world' find themselves under pressure to conform to androcentric 1/0 reasoning. Yet identification with middle-class European norms can lead both indigenous and feminist liberals into misbegotten policy manoeuvres. This was seen at the 1994 UN International Conference on Population and Development in Cairo, where the economic causes of land degradation and poverty were put aside, in favour of a twenty-year policy consensus on 'population control'. Masses of women in Brazil have been sterilized through international programmes, and one Indian National Family Health survey records the average age of women subjected to the practice in that country as twenty-six years.

As Patricia Hynes, a pioneering US ecofeminist argued, it is well time for taking a harder look at global resource distribution and consumption and *Taking Population Out of the Equation* altogether.[6] Then, at the UN Fourth Decade Conference on Women in Beijing 1995, liberal feminists were shocked by what they heard. Lynette Dumble, an expert on Depo-Provera, Norplant and RU486, exposed the misogynist and genocidal medical paradigm that drives debate over the population control:

> long-acting contraceptives that at one extreme may blind women by increasing the pressure within their brain cavities . . . vaccines that render women infertile by creating auto-immune disease; mass sterilisation camps where women die on a regular basis; medical experiments with hormones and an array of other chemicals that disrupt women's fertility or terminate their pregnancies with little or no concern for the acute ill-effects, let alone the chronic future morbidity.[7]

Typifying the interlock of a profitable corporate sector and an ostensibly independent international body such as the Population Council, Upjohn Pharmaceuticals was donating US patent rights for Depo-Provera to that Council, whose bureaucrats in return ensured an ongoing market for the product. Writing in 1995, Dumble pointed out that the contraceptive toxin, Depo-Provera, is surreptitiously being used on disabled, Black and Hispanic women in the United States and on Aboriginal and immigrant women in Australia. Once again, 'women and natives' would be targeted as 'vermin' and now, in the twenty-first century, the population question is still at the front of

Big Pharma's concerns. What Dumble concluded a quarter century ago still applies. The pharmaceutical industry is assembling an 'armour of pesticidal, or more specifically femicidal weapons . . . I have [even] seen an Internet suggestion that population expansion could be more rapidly halted from the use of a genetically engineered virus'.[8] The colonial assault on nature, land, animal and human bodies is energized by a desire whose roots reach far deeper than empire or even profiteering. In the early days of Australian settlement, horseback raids by lusty gangs of up to 100 men 'sterilized' the new country by chasing down every living thing from tribal families to wallabies and emus.[9]

The colonial history of Africa goes back much further than settlement of Oceania; but it too is moving into modernity through public relations techniques of 'partnership, capacity building, and harmonisation'. In each region, resolution of 'the Land question' remains the cutting edge of the World Bank-IMF activities whereby men of the Global North unravel uniquely communal relations of social reproduction. The pattern appeared in New Guinea and in Vanuatu, with registration of custom lands being a first step towards privatization and thus negotiation with outside investors. As happened in Europe centuries before, once land is valued, a rising urban middle class is able to leverage itself upwards with an eye for a well-paying operation. But dislocated families are left with nothing but the labour, or less, the organs of their bodies to sell for livelihood. In Africa, Asia and South America, the destructive cycle of the World Bank loans forces small holders into cash cropping to meet debt repayments. The strife that follows is usually attributed to 'religious wars' in the international press, but George Caffentzis tells the early stages of such capitalization this way:

> starvation, mass forced migrations, wars of extirpation and plagues are, of course, the violent symptoms of the most fundamental *liberation of labour power* which is known as *primitive accumulation* . . . [this] involves also the expropriation of the body, of sexual and reproductive powers, in so far as they are a means for the accumulation of labour.[10]

The social disruption from enforced enclosures has been especially hard on women; African infant mortality has risen and life expectancy declined. The advent of AIDS swelled the reserves of cheap labour pushing down its price to desperate levels. In Nigeria as elsewhere, harsh Structural Adjustment Programmes (SAPs) would be introduced to assist national debt repayments, cancelling health and welfare and leaving women especially dependent on relatives. Traditional practices like female genital mutilation, a cause of sterility, remain low priority. Caffentzis's judgement is that

> The second success of the debt crisis is in [mastering] the African body, a male/female body of mythic dimensions in the imagination of economic analysts. For the economic consequences activated by the debt crisis and SAPS have given legitimacy to their attempt to control African fertility... by 1984 A. W. Clausen (then president of the WB) ... called for a 'social contract' between African governments and African parents.[11]

Silvia Federici has noted how the Nigerian government encouraged by the World Bank was prepared to tax women who procreate beyond 'the optimal level'; at the same time, it subsidized wealthy transnational oil cartels who pollute arable land.[12] The World Bank-IMF structural adjustment policies also contributed to the crisis in Rwanda. And despite a UN embargo, a British company, Mil Tec Corp, cashed in on the genocide, supplying Eastern European-made rifles, grenades and mortar bombs to the Rwanda government secretly via Zaire. The deal brought home a queenly profit of $6 million.[13]

BioColonialism

Struggles for 'difference', that is, for cultural autonomy and natural biodiversity, come together over the matter of Trade Related Aspects of Intellectual Property Rights (TRIPS). In Australia, Henrietta Fourmile from Cape York would point out that the continent's biodiversity consists of some 475,000 plant and animal species.[14] And further, that the system of totem identification within Aboriginal Customary Law is the oldest surviving system of usage rights. These Common Law rights are recognized in the Biodiversity Chapter of *Agenda 21* and in the Native Title Act, Section 212. But such provisions are little help, given biopiracy by Big Pharma and nurture of the biotechnology industry by the Australian Department of Foreign Affairs and Trade. Without due acknowledgement researchers from the Commonwealth Scientific and Industrial Research Organization continue to raid knowledge of biodiversity built up over centuries of Aboriginal groundwork. Hand in hand with entrepreneurial bioprospectors, scientists rake through this genetic heritage, 'reserving' what they want in seed or gene banks. Fourmile notes that concurvine, a plant with potential to cure AIDS, can draw millions in royalty dollars, but Aboriginal people see none of it. Although the Biodiversity Convention allows for 'farmers rights', so far nothing has been paid out in exchange for use of genetically cultured stock. Part of the reason for this may be infiltration of the UN Food and Agriculture Organization (FAO) by the international business pressure group Consultative Group on International Agricultural Research (CGIAR).

Given how the cards are stacked under transnational capitalism, the people of India achieved a great victory in a case involving their native neem tree. The United States Patent and Trademark Office had granted W. R. Grace a pesticidal patent derived from the plant. So in 1996, the United States took legal action against India claiming local use of the neem to be a violation of Clause 301 of the US Trade Act. Company lawyers argued that the new product was 'a synthetic compound', not pirated knowledge. However, in a move led by the Foundation for Economic Trends, International Federation of Organic Agriculture, Third World Network and Navdanya, some 200 organizations successfully overturned the ruling and restored rights over the plant to India. The case was a landmark in intellectual property legislation administered under the WTO.[15] Communities in India now lead the way in making inventories and seed banks using traditional methods.[16] Meanwhile, arguing from the precautionary principle, activists at the Pacific Concerns Resource Centre in Fiji have pushed for nothing less than a moratorium on genetic engineering altogether, reserving the South Pacific as a 'patent-free zone'.

However, DNA from the blood, tissue and hair of Aboriginal and Torres Strait Islander communities has already been 'tapped' and 'banked' as part of the Human Genome Research Project. That US research programme is funded by the National Institutes for Health and the Defense Department, both having an interest in biological warfare. In this context, expansive commercial phrases like 'the common heritage of Man' reveal profound ignorance, if not complicity. First Nations may appeal to the UN Convention on Human Rights and Childrens Rights, the Draft Declaration on Rights of Indigenous Peoples', the ILO Convention and the International Convention on Civil and Political Rights, Article 29. But to echo Mick Dodson's words: 'basically the existing legal system cannot embrace what it needs to define.'[17] Nor, it seems, can the patriarchal-colonial-capitalist imperium define what it needs to embrace – that is, an identification with Life-on-Earth. The Northern Territory Conservation Commission shares few of these concerns. Exercising the Eurocentric penchant for instrumental rationality, state bureaucrats simply objectify biological resources as 'organisms or parts thereof, with actual or potential value for humanity'. And more: patenting is described as 'a way of organising order out of chaos'.[18] To foster R&D, the Territory government very early on brokered biotechnology deals between AMRAD pharmaceuticals and the Tiwi people; with the Northern Land Council; and with its elected self for an undisclosed consideration. Any notion of government by the people for the people appears to have become an anachronism. The worldwide strategic plan of mining and agro-industry, updated daily now by electronic conferencing, has turned governments into mere stewards of private enterprise.

It was the 1995 Jakarta meeting on the Biodiversity Convention that set the seal on corporate patenting of live DNA from plants, animals and human beings. A protocol on biosafety had been consistently obstructed by the US vassal states on its behalf. Like the mining giants, biomanufacturers favour 'voluntary regulation'. But unlike mining pollution, which remains local, there is no way of determining how far genetically engineered organisms will spread – nor what their effects on people and habitat might be.[19] Meanwhile, expensive public relations firms retained by the genetic engineering industry sell the new 'science' across the global media. But what is going on is very half-baked science, with no thought to consequences for ecological systems or human evolution. Consider too, the unspoken social costs of using GISs to preserve indigenous knowledge. The digital methodology instantly subverts indigenous ways of knowing, fine-tuned as these are by sensuous interaction with what they call 'country'. Again, when an oral tradition is extracted from its generational context, what impact will that have on the social well-being of a community where Elders are pivotal to social integration? Surely, the very translation of indigenous knowledges into 'resource speak' betrays their cultural meanings.

Beyond this, if GIS data is available only to those with computer access, is that democratic? Who will profit from the sale of local knowledges on CD-ROM or by circulation through IT sites? Will GIS play into the hands of corporate interests, currently centralizing the global food industry, and what hope then for self-sufficiency? In a democratic, non-racialized, non-speciesist world, people would seek to understand how customary classification systems are put together and to learn from eco-sufficient indigenous skills.

Nonsensical Law

In the long run, to base 'value' on markets is to adopt the founding premise of capital, where only what is 'improved by Man' – the commodity – has worth. Even well-meaning radical groups like Cultural Survival Enterprises promote indigenous forest products for international sale. One Australian indigenous Elder, Professor Marcia Langton, has dismissed the ecology movement as 'a barometer of colonial anxiety'. And while there is an element of truth in this, the claim overlooks a world of difference between how business leaders at the World Economic Forum think and how most ecological activists think. True, there are affluent and self-serving environmentalists inside the International Union for Conservation of Nature (IUCN) or the leisure-oriented US Wilderness Movement or the North Queensland minority who dub themselves Sanctuary Protectors. But mainstream groups, like the Australian Conservation Foundation, the World Wildlife Fund and Greenpeace, have also worked alongside Aboriginal people to refine Land Rights policy, Pay the Rent, and get better provisions on hunting, fishing and parks. Closing the Gap between rich and poor nations will depend on the Global North scaling down taken-for-granted levels of resource use, while acknowledging that many indigenous economies already demonstrate self-sufficiency.[20] They model a stable interchange with habitat, use low-impact technologies and work few hours a day while giving energies to social bonds, ceremony and art.

Ecologists taking a lesson from Aboriginal cultures might discover how to devise low-demand, low-impact ways of provisioning where sustainability and social equity can go together. In Australia, the colonial *terra nullius* myth was a self-serving characterization of the continent as basically empty, occupied by a few 'primitive hunter-gatherers'. However, a re-reading of letters, diaries and sketches left by European explorers and early settlers shows there was an appreciation of indigenous cultivation practices and seed trading. And following the discovery of grindstones dating back 30,000 years at Cuddie Springs near Walgett, New South Wales, Elder Bruce Pascoe reflected:

> This makes these people the world's oldest bakers by almost 15,000 years, as the Egyptians, the next earliest, didn't bake until 17,000 BC. Other peoples ground tubers to extract starch, but it seems that Aboriginal people were the first to discover the alchemy of baking bread from the flour of grass seeds.[21]

Similar tools have been found at Kakadu in the Northern Territory. Stone blade knives with fur-covered handles were used. Fire was applied selectively as a tool in hunting and also to temper weather conditions. The early record speaks of tribal women scattering and watering seed, harvesting, threshing and stacking the surplus in haycocks. There is evidence of seed domestication, irrigation and aquaculture using skilfully designed fish traps. Once Europeans arrived, alien animal stocks devoured the oat grass, yams, nardoo and imported hooved animals compacted soils such that indigenous communities lost not only dwelling sites but sustainable subsistence economies too.

Seasonally, the colonial gaze leaves commerce behind and turns to wilderness as pure and untouched land; indigenous peoples are romanticized on tourist posters. Marcia Langton has been right to question this, as well as the classic Australian pastoral idyll of the outback home, sustained at heavy cost to Others held back by the 'vermin fence'. Modern wilderness 'husbanding of virgin lands' through national parks further extends the conquest, displacing indigenous skills and livelihood. Transnational corporations (TNCs) used the foundational 1992 Earth Summit to push openly for more global 'enclosures', even privatization of national parks; but by definition 'the wild' must be what escapes control. In fact, the wild speaks potentials to rediscover beyond the *Androcene*. For by an ecocentric ethic, country is never vacant, as in *terra nullius*, but an intractable subject in its own right. This notion already exists in customary law. Movements for social change beyond virtual politics can make good use of the wilderness idea, by rejecting the 1/0 projection of Nature as 'out there' and independent of human selves. Understanding just how to do that is important for *Androcene* others like workers, women and youth. For without healing the Humanity/Nature, Civilized/Native split – and the denials that such psychological dissociation thrives on – contemporary political efforts will simply fall before the objectifying modernity of measurement and growth.

The Australian Land Rights legislation embodied in the 1993 Mabo case should have dealt with the attitude of *terra nullius*. However, as anthropologist Jon Altman points out, corporate interests actually gained from it, as the law intentionally refused to confer free, prior and informed consent rights over mineral extraction to native title groups. This was despite the existence of an earlier veto in the *Aboriginal Land Rights (Northern Territory) Act* 1976 (Cth).[22] To First Nation peoples, coloniality remains arbitrary, logically absurd and inhumane: and so the contradictions of history roll forward into the tragedy of Juukan Gorge.

> In the days leading up to National Sorry Day and the opening of National Reconciliation Week in May 2020, the mining giant Rio Tinto destroyed two Juukan Gorge sites belonging to the Puuntu Kinti Kurrama and Pinikura (PKKP) peoples. These sites were said to contain some of the earliest human expressions of symbolic meaning on the planet. They are the only known inland sites showing evidence of continual human occupation back into the last ice age, 46,000 years or more.[23]

After a perfunctory apology and pause in mining activities, the company and others – namely Fortescue Metals, Australian Potash and China Shenhua – have continued apace in the Pilbara region. In Western Australia, some 400 applications to mine in culturally sensitive sites have received government approval. At Darwin, in the Northen Territory, the Beetaloo Basin is planned to source new liquid natural gas exports as well as a petrochemical plant producing plastics, fertilizers and paints. Besides greenhouse gas emissions, air-born particulates are estimated to rise by 500 per cent. In this material context, the well-meaning public advocacy of Closing the Gap between Black and white Australians has a hollow ring.[24]

Over the years, elected consultative bodies like the Aboriginal and Torres Strait Islander Commission (ATSIC) have come and gone, although a number of elected indigenous members serve in the Australian parliament. In 2023, a new Labor government in Canberra took up the indigenous community's 2017 Uluru Statement from the Heart – a call for recognition as a Voice in the country's Constitution.[25] However, the attempt to enact The Voice by means of a nationwide Referendum was unsuccessful. Some international observers judged this an effect of Australian settler racism, but the fallout was more complex. In fact, both Yes and No positions on The Voice were internally divided among themselves, with each side divided three ways. Proponents of The Voice included Black academics and lawyers, some with old links to the mining industry; Torres Strait Islander communities; and a would-be fair-minded Middle Australia. Opponents of The Voice included the federal opposition being a right-wing Liberal-National Coalition; Black left-wing intellectuals seeking Truth Telling and a Treaty; and neglected Outback communities, alienated by the whole process. Indigenous activist Celeste Liddell surmised that not only was the government's implementation of The Voice idea poorly conceived, it was also an exercise in 'wallpapering'.[26] The nationalist policy framing was anachronistic in a decade when the 'development idea' is giving way to bioregionalism and an appreciation of cultural diversity. A local expression of this emerging global pluralism is clear in Tyson Yunkaporta's account of one Elder's wit-sharp 'bush lawyer' deconstruction of the colonial legal system. By the judgement of this Uncle, indigenous customary law is still in place, and this is seen in how many First Nation people disregard their state-authorized birth certificate. In their eyes, the document was issued under the logic of *terra nullius*, so it is merely a fictional identity – for 'sovereignty was never ceded'.[27]

4

Nuclear Risks

Voices for Life-on-Earth

On 11 March 2011, the Fukushima nuclear electricity plant in Japan was hit by a powerful earthquake and tsunami. An undetermined area of land remains uninhabitable; thousands of people are trying not to breathe, touch, eat or drink the toxic levels of radiation in their environment. It is believed that BHP Billiton's Olympic Dam and Rio Tinto's Ranger mine exported uranium from Australia to this reactor. Now confusion and anger, sickness and disability will mark many Japanese lives for years to come.[1]

> Over 80,000 people have been forced to abandon their homes. Thousands of people are now without a livelihood or the hope, in the near future, of rescuing one. Compensation claims are certain to be well over $100 billion; rebuilding of infrastructure and housing will cost at least $200 billion. Then there's the cost of clearing over 20 million tonnes of rubbish, some of it radioactive, and the cost of securing and decommissioning the stricken reactors over the coming decades. Add to this the relocation of people and factories and the settling of injury and health issues, and the cost of this disaster will be in the neighbourhood of $450 billion, just a little under 10% of Japan's GDP. There are an estimated 1,000 corpses too radioactive to retrieve. Even when they are, who will cremate or bury them, and where?[2]

Denialism

Fukushima was a civilian incident, but nuclear power and military weapons are joined in the global production system. After the Second World War, occupied Japan would enter an economic boom as chemical weapons were converted into pesticides for farms and nuclear know-how turned into power for cities.[3] I, myself, started thinking seriously about nuclear radiation in 1976 after hearing a talk by the Australian paediatrician Dr Helen Caldicott. As a mother and worker in Aboriginal communities at the time, within days, I was helping set up a Sydney branch of the Movement Against Uranium Mining and within months we had 100,000 people marching down George Street. For a while, the Australian Labor Party spoke with the people's voice, but its political will gave way eventually to the mining lobby. In the United States, Caldicott's efforts at public education were also targeted by the energy cartel's media outlets. As she pointed out in a letter to the *New York Times*, the nuclear industry can only survive by misleading the public.[4] Physicists talk of a 'permissible dose' of radiation, but biologists know there is no such thing. The fact is that radiation damage in the body takes time to reveal itself.

Nuclear denialism takes place in private and public sectors. Installation accidents at Windscale in Cumberland, UK, 1957, and at Three Mile Island in Pennsylvania, United States, 1979, were largely 'contained' by public relations expertise. Following the meltdown at Chernobyl, USSR, 1986, an embarrassed Soviet government failed to guide its citizens with health advice. Caldicott observes that today, both Belarus and Ukraine have group homes full of deformed children. After the Chernobyl cloud crossed Turkey, leaders were so determined not to panic 'the people' that relevant information was censored. Doctors who helped mothers terminate pregnancies were jailed, and journalists who tried to report this were jailed too. In terms of cancer outcomes, Peter Karamoskos, a nuclear radiologist, and medical doctor Jim Green offer the following assessment of Chernobyl:

> The International Atomic Energy Agency estimates a total collective dose of 600,000 Sieverts over 50 years from Chernobyl fallout. A standard risk assessment from the International Commission on Radiological Protection is 0.05 fatal cancers per Sievert. Multiply those figures and we get an estimated 30,000 fatal cancers.

But they go on to add that:

> In circumstances where people are exposed to low-level radiation, studies are unlikely to be able to demonstrate a statistically significant increase in cancer rates. This is because of the 'statistical noise' in the form of widespread cancer incidence from many causes, the longer latency period for some cancers, limited data on disease incidence, and various other data gaps and methodological difficulties.[5]

Formulae for calculating nuclear casualties vary, but the problem of denial is a constant. Australian firms, such as Toro Energy, Uranium One and Heathgate Resources, are known to have sponsored lecture tours by industry scientists who dismiss public concerns about radiation.[6] Since the Fukushima meltdown, Japanese citizens have become increasingly disturbed by an absence of transparency from both the Tokyo Electric Power Company (TEPCO) and government officials. And neither the World Health Organization nor the International Atomic Energy Agency has provided women with information about radiation exposure effects on their reproductive function.[7] If anything, disinformation is the order of the day. A *Wall Street Journal* article quotes Genichiro Wakabayashi from Kinki University's atomic energy research institute, claiming that wearing masks or staying indoors during summer will harm children more than radiation will.[8] So too, Japanese people have been encouraged to support their country by eating local produce. Yet as Roger Pulvers tells it:

> No one knows how badly the sea around Fukushima has been contaminated, and we are only beginning to assess the effect that radiation has had on the land. Several hundred kilograms of tainted beef from Fukushima have been sold to markets as far away as Kagoshima on the southern island of Kyushu. This beef has registered up to 2,300 becquerels of radioactive caesium per kilo, more than five times the government-set safety limit. 648 head of cattle in Fukushima, Yamagata and Niigata Prefectures have eaten contaminated straw. It has been shown that the feed itself contained up to 57,000 becquerels of radioactive material per kilogram.[9]

The self-interest of those who deny nuclear risk is both capitalist or economic and masculinist or cultural.[10] Psychological denial protects a structural hierarchy of wealth, power and bonding opportunities between men. But near the lower rungs of this narrow ladder of rewards stand youth, indigenous peoples and women – the Others of neoliberalism and its hegemonic masculinity. These others exist in direct contradiction to the military-industrial complex, and they each bring complementary insights and skills to its political transformation. However, my focus in this chapter is on women, mothers and housewives, many of whom are also indigenous, giving double strength to their political work. People whose labour sustains human bodies and links to natural habitat prioritize social reproduction over economic production. This observation gives rise to a distinct political analysis known as ecofeminism. It emerged fifty years ago, from thinkers and activists on every continent, and the nuclear question was central to it.[11] What is unique about women's resurgence in ecological struggle is how they combined it with their self-understanding as 'women'. Their focus on pollution was and remains both inner and outer, personal and political. Women demeaned by men's objectification of their humanity as 'femininity' feel a need to purify and rebuild a self-identity on their own terms. Ecofeminists, as they call themselves, reject what they see as 3,000 years of maldevelopment in the social construction of sexed and gendered relations. Their political activity, based on a mode of being grounded in identification with the primal parent rather than objectification, goes hand in hand with attention to psychological growth in mutually supportive consciousness-raising sessions. This revolutionary strategy is a profound existential commitment. However, women have come to be disappointed to find so few environmentalist brothers entering into a parallel reflection on selfhood under the predatory model.

Women's Collectives

What follows is a short review of the formative years of this radical resistance and the rise of 'management' environmentalism with its cultivation of liberal feminists, before coming home again to the situation in Japan.

In the United States, as far back as 1962, lawsuits against the corporate world were coming out of the kitchens of mothers and grandmothers – *Mary Hays v Consolidated Edison, Rose Gaffney v Pacific Gas, Jeannie Honicker v Nuclear Regulatory Commission, Kay Drey v Dresden Nuclear Power Plant, Dolly Weinhold v Nuclear Regulatory Commission at Seabrook*.[12] Japanese women were also foot soldiers in campaigns against local pollution. One Ishimure Michiko founded the Citizens' Congress on Minamata Disease Countermeasures in 1968. Other women set up the path-breaking producer-consumer cooperative known as the Seikatsu Club – which economic model would grow to some 200,000 or more members.[13] Parisian writer Francoise d'Eaubonne's book, *Le feminisme ou la mort*, and the US Democratic Socialist Rosemary Ruether's *New Woman: New Earth* gave early intellectual impetus to ecofeminism. A conjectural history of the self-deforming practices of mastery was drawn. If the Greek word *'oikos'* was etymological root of both ecology and economics, the latter had lost its way. In 1974, the unquiet death occurred of whistleblower Karen Silkwood, a unionist at Kerr-McGee's Oklahoma plutonium processing factory. In 1975, women blockaded land clearing for construction of a nuclear reactor at Wyhl in Germany. More than economic loss of vineyards, they said, it was a matter of 'our human-being-in-nature'. By 1976, in Australia, women Friends of the Earth in Brisbane were conferencing on women and ecology and some taking a coordinating role in the new Movement Against Uranium Mining. Even the mainstream women's magazines were printing pieces on women and the anti-nuclear issue. In 1977, a consciousness-raising group Women of All Red Nations (WARN) emerged among tribal Indians in South Dakota. They were especially worried about weapons tests, aborted and deformed babies, leukaemia and involuntary sterilization among their people.

Women circulated articles on artificial needs and consumerism, animal exploitation for cosmetic manufacture, recycling, indigenous health and, of course, uranium.[14] Separatist anti-nuclear groups were established in Australia – Women Against Nuclear Energy (WANE) in the Eastern states and a Feminist Anti-Nuclear Group (FANG) over in the West. Women's ecology collectives started up in Paris, Hamburg and Copenhagen, and ads for feminist organic farming communes appeared on every noticeboard. Susan Griffin's *Woman and Nature: The Roaring Inside Her* was published in 1978.

Elizabeth Dodson Gray's *Green Paradise Lost* followed in 1979. Each author in her own way described the self-alienation of the androcentric ego construct; the obsession with control of othered peoples; the fascination with militarism; and its counterpart in instrumental logic and scientific calculation. Women wanted nothing less than a new language, reintegrating reason and passion.[15] In the late 1970s, the US League of Women Voters began lobbying for a moratorium on nuclear plant construction licences; the YWCA initiated an anti-nuclear education campaign; while the National Organisation of Women (NOW) instituted a National Day of Mourning for Silkwood. A further group – Dykes Opposed to Nuclear Technology (DONT) – organized a New York conference on the energy crisis, a patriarchally generated pseudo-problem, and a Women and Technology Conference was held in Montana the same year. Delphine Brox-Brochot of the Bremen Greens called for an end to high-tech aggrandisement while millions around the world still starve. Everywhere in the so-called 'developed world', women's political lobbies and protests over effects on workers and children of pesticides and herbicides, of formaldehyde in furniture covers and insulation, of carcinogenic nitrate preservatives in foods, of lead glazes on china were gaining momentum. But there was a weary road ahead, to quote Joyce Cheney: 'I am annoyed that I feel forced to deal with the mess the boys have made of the earth. It is a hard enough struggle to survive and to build and maintain a life-affirming culture.'[16]

In 1980, a collective called Women Opposed to Nuclear Technology (WONT) organized a Women and Anti-Nuclear Conference in Nottingham, UK. Women in Solar Energy (WISE) began meeting in Amherst, Massachusetts, and Ynestra King mounted the first Women and Life-on-Earth Conference. By November 1981, a 2,000-strong body of women marched on the US capital, symbolically encircling the Pentagon. By now, Helen Caldicott, president of Physicians for Social Responsibility, had started a Women's Party for Survival in the United States, with some fifty state and local chapters. This was subsequently broadened to become Americans for Nuclear Disarmament. In India, the *Manushi* collective published their influential piece 'Drought: God Sent or Man Made Disaster?'[17] Carolyn Merchant's classic thesis *The Death of Nature: Women, Ecology and the Scientific Revolution* began to make itself felt in academic circles from this time on.[18] By the mid-1980s, the following networks were operating in the United States: Lesbians United in Non-Nuclear Action (LUNA) v Seabrook Reactor; Church Women United; Feminists to Save the Earth; Feminist Resources on Energy and Ecology; Dykes Opposed to Nuclear Technology (DONT) v Three Mile Island and Columbia's TRIGA Reactor; Women for Environmental Health demonstrating in Wall Street; Mothers and Future Mothers Against Radiation v Pacific Gas and Electricity; Women Against Nuclear Development (WAND); Spinsters Opposed to Nuclear Genocide (SONG); and Dykes Against Nukes Concerned with Energy (DANCE) v United Technology. Women's environmental conferences were held at Sonoma and San Diego State universities.

In Japan, a *kamakazi* encampment of grandmothers known as the Shibokusa women was running continual guerilla disruptions on a military arsenal near Mt Fuji, while a further 2,500 women marched on Tokyo in the cause of world peace.[19] By 1981, WONT had grown into a string of non-violent direct action cells around the UK; many began what would become the perennial encirclement of Greenham Common missile base; and in Germany 3,000 women were demonstrating at Ramstein NATO base. In Australia, Margaret Morgan drew together a rural anti-nuclear organization at Albury, and the *Sun Herald* newspaper was reporting on Labor Party and Democrat women's decisive interparty policy stand against lifting bans on uranium mining. In 1983, a new collective, Women's Action Against Global Violence, was encamped at Lucas Heights Atomic Energy Establishment near Sydney. This was followed by a protest in the desert with Aboriginal men and women outside the secret US reconnaissance station at Pine Gap. A first ecofeminist anthology *Reclaim the Earth* was brought out by Leonie Caldecott and Stephanie Leland.[20] An Environment, Ethics and Ecology Conference in Canberra opened up a debate between women ecofeminists and not so sex/gender aware Deep Ecologists.[21] British elections saw a combined Women for Life-on-Earth & Ecology Party ticket; and a year later, ecofeminist Petra Kelly led Die Grunen into the Bundestag. Kelly's passionate biography, translated as *Fighting for Hope,* told how her anti-nuclear politics began as she watched her young sister die of leukaemia.[22]

From Victims to Leaders

The soviet reactor accident at Chernobyl in 1986 alerted women to the lack of accountability in capitalism and socialism alike. Across Germany and Eastern Europe, a 'birth strike' expressed outrage, as governments from Turkey to France suppressed vital facts about environmental radiation levels for fear of damaging national economies. Sami people to the north of Scandinavia met official lies about post-Chernobyl radiation with a firm resolve for land rights. From the other side of the Earth in Australia, Joan Wingfield of the Kolkatha tribe flew from the Maralinga site of 1950s–60s British bomb tests to address an International Atomic Energy Agency conference in Vienna. In Germany, Maria Mies published *Patriarchy and Accumulation on a World Scale*, the first substantial socialist ecofeminist statement.[23] A more New Age rejection of high-tech 'progress' was US bioregionalist Chellis Glendinning's *Waking Up in the Nuclear Age*. In 1987, Darlene Keju Johnson from the Marshall Islands and Lorena Pedro from Belau, both Women Working for a Nuclear Free and Independent Pacific, went public about the jelly fish babies born to islander women and cancers in ocean communities following US atom tests.[24] The First International Ecofeminist Conference was held in 1987 on campus at the University of Southern California. North, South, East and West, women's commitment to Life-on-Earth now spanned the nuclear threat, reproductive technologies, toxic chemicals, indigenous autonomy, genetic engineering, water conservation and animal exploitation. Depleted uranium would become a focus with the Balkan and Middle East wars. Women's International League for Peace and Freedom (WILPF), Code Pink, Madre, PeaceWomen and the World March of Women continue to pursue many of these concerns.

It is now three generations since ecofeminists came to politics; the movement continues to grow in experience, cross-cultural networks and theoretical sophistication.[25] Debates over sex/gender literacy in environmental ethics or ecosocialist formulations have become standard fare for university courses, academic journals and publishing houses. International initiatives by Vandana Shiva have even been recognized with an Alternative Nobel Prize.[26] Ecofeminism is at once an autonomist socialism, an ecology, a decolonial movement and a case for respecting women's initiatives in designing 'another world'. This said, ecofeminist work has been affected by changes in the political character of both feminism and environmentalism. Occasionally, one-dimensional thinkers blind to the depth and complexity of women's ecopolitical renaissance judged it to be little more than a public extension of the housewife role. Articles appeared with patronizing and demeaning titles like 'Still Fooling with Mother Nature' and 'Calling Ecofeminism Back to Politics'.[27] But a glance at the now extensive literature of ecofeminism shows its reach from epistemology to economics. My sense is that the establishment had become uneasy about this radicalism quite early on, because as women were writing their herstory, transnational corporations stepped up proactive measures – structural and ideological – for taking global control of the environmental agenda.

In the structural domain, the principle of neoliberal competitiveness would be legally embedded in international treaties and bureaucratic agencies like the UN. First the 1987 Brundtland Commission routinized a materially contradictory policy of growth with 'trickle down benefits' for sustainability. Then the 1992 Rio Earth Summit leveraged this up, setting the politics of Biodiversity and Climate Change Conventions in motion.[28] Soon the Kyoto Protocol and a rolling international agenda of Conference of the Parties (COP) meetings would have movement activists running to keep up with the newly institutionalized ecomodernist discourse of environmental management, while the public was carefully marginalized and disempowered by the academic complexities of 'risk analysis' and 'biosecurity'. The globally orchestrated politics of liberal environmentalism enlisted the UN, private foundation and government sponsorship of special women's ecology organizations to 'mainstream' women's views in international policy. Women's 'citizenship' became the new liberal mantra. Women's Environment and Development Organization (WEDO), founded by veteran US Congresswoman Bella Abzug in the early 1990s, played a big role in this. Thus, at the UN Framework Convention on Climate Change COP13 in Bali, December 2007, Women in Europe for a Common Future are found hard-pressed keeping nuclear power out of the Clean Development Mechanism. So depth analysis of hegemonic masculinity gives way to ironing out its incoherencies.

Interminable international environmental meetings focus on women as 'victims' or objects of natural disaster, and women who play the liberal feminist card to this policy are rewarded as 'professionals' for not rocking the androcentric boat too much. There is no place for an ecofeminist diagnosis of the cultural context of such 'crises'. Nor is the knowledge of indigenous women from say Oceania acceptable as an existing model of low-carbon provisioning. Instead, the German Federal Ministry for the Environment, Nature Conservation and Nuclear Safety would draft women from the Global South into 'capacity-building' workshops for 'climate adaptation and mitigation'. While such neoliberal operations are ostensibly about 'justice and sustainability', the orientation is always framed by business as usual. In the ideological domain, 'management environmentalism' relies on several techniques for the pacification of citizens and governments. Public relations firms are employed to 'greenwash' or minimize local damage from capitalist industrial enterprises.[29] Again, the packaging of ecology as a media commodity thins out the reporting of grassroots voices in favour of a few colourful and iconic feminist 'personalities'. A further silencing of ecofeminist politics has occurred as a result of public reliance on the internet as chief recorder of radical movements – since 90 per cent of web-based material is selected and posted by men – radical youth notwithstanding. A final ideological assault on women's ecological struggles has come through the universities. In the 1990s, as Left analysis was overtaken by a new field of cultural studies, many women students took to the deconstructive study of political texts, an innocent but elitist move, leaving the concerns of threatened communities far behind.[30]

Enough Looting!

While the institutions of Eurocentric globalization insured themselves against critique from within, peoples at the geographic periphery had been addressing the 500th year of Columbus. Then, at the 1992 Earth Summit in Rio de Janeiro, grassroots environmental politics would implode, taking a distinctly decolonial turn. The articulation of this perspective by South American activists is very rich. As anti-nuclear activists from the Arrernte, Tuareg nomads, and Acoma

Pueblo, spoke truth to power in Washington, a First Continental Summit of Indigenous Women in Peru produced a Manifesto in the cause of all life. The preamble to this 2009 document shows the women weaving together a seamless politics of sex/gender, ethnicity, class and species justice.

> We are the carriers, conduits of our cultural and genetic make-up; we gestate and brood life; together with men, we are the axis of the family unit and society. We join our wombs to our mother earth's womb to give birth to new times in this Latin American continent where in many countries millions of people, impoverished by the neo-liberal system, raise their voices to say ENOUGH to oppression, exploitation and the looting of our wealth. We therefore join in the liberation struggles taking place throughout our continent.[31]

In short, as the *Mujeres Creando* of La Paz have it, there can be no decolonization without depatriarchalization.

In Bolivia, this strongly integrative indigenous politics opened into The People's Alternative Climate Summit at Cochabamba, April 2010, advancing a substantive economy based on the principle of 'living well', to replace the death-risking formal economy of the mega-machine.[32] In 2011, the circle closes with Vandana Shiva and Maude Barlow seeking UN ratification of a Declaration of the Rights of Mother Earth: 'affirming that to guarantee human rights it is necessary to recognise and defend the rights of Mother Earth and all beings in her and that there are existing cultures, practices and laws that do so'.[33] In the current crisis of global warming, the international nuclear industry presents itself as 'a clean, green, alternative' to fossil fuel-based power generation. But not only is it a threat to all natural processes, the engineering of installation components and their daily operation draws massive amounts of electric power. Nevertheless, Japan's ruling class with US corporate partners aims to put nuclear power back on track with more science and better 'technocratic management', even as Silvia Federici and George Caffentzis point out:

> the damaged nuclear reactors can hardly be blamed on the lack of capitalist development. On the contrary, they are the clearest evidence that high tech capitalism does not protect us against catastrophes, and it only intensifies their threat to human life while blocking any escape route.[34]

It is not rational to pursue a fantasy of ecomodernisation by means of this arsenal. The Fukushima meltdown may be a bonanza for reconstruction companies like Haliburton once they're done in Iraq, but the revolving door of men in suits knows well that 'business is merely war by other means'.

The crisis of Fukushima should have become a political turning point. Certainly Japanese women and men have pioneered nuclear resistance. I think of the late Women and Life-on-Earth activist, Satomi Oba, president of Plutonium Action, Hiroshima.[35] And the perennial warnings of Kenji Higuchi, much sought after for the lecture circuit now.[36] Hisae Ogawa and others in the international ecofeminist peace organization Code Pink are working all over Japan. Friends of the Earth is attending the special needs of women and children, demanding wider evacuation zones and sackings in high places. Greenpeace is encouraging the public to mobilize, and in the months since 11 March 2011, mass demonstrations rolled across Japan urging the end of nuclear power. Suddenly politicized, angry mothers and housewives took to the streets in their thousands. This nuclear disaster re-energized international opposition to the industry and women's organizations remain highly focused. The Asian Rural Women's Coalition meeting in Chennai has condemned plans for nuclear power plants in India, Burma, Thailand, Indonesia and the Philippines. The Gender-CC Network contests nuclear power through its regular climate change campaigning.[37] In the United States, the National Organization of Women (NOW) and the United Farm Workers look into possible bioaccumulation of radioactive caesium from Japan in California cows' milk.[38] In Australia, indigenous women continue fighting the government's proposed nuclear waste site on their land at Muckaty, Northern Territory.

The Asia Pacific Forum on Women, Law and Development, an NGO with consultative status to the UN, would write to the prime minister of Japan, observing the unique vulnerability of women in post-disaster situations – as objects of violence, as part-time employed and as those doing most of the country's care work. They noted that only one woman sits among the sixteen members of the Reconstruction Design Council. They referred the prime minister to Japan's obligations under the United Nation's Committee on the Elimination of Discrimination Against Women (CEDAW). They urged that 'gender' disaggregated statistics be collected to prepare 'gender-specific' budgets.

And the letter requests the Japanese government to exercise accountability by consulting with local women's organizations and promoting women's participation as planners and decision-makers at prefecture, municipal and town council levels.[39] How can a country call itself a democracy when it does not give women equal seats on its Reconstruction Design Council? Could the achievement of this liberal feminist objective actually turn Japan around? Like the affirmative action for women at big international environment meetings, it would simply paper over an unjust and unsustainable order. An ecofeminist politics is essential to expose and neutralize the *Androcene* that let Fukushima happen. A balanced committee is one thing, but it is even more essential to redefine its 'terms of reference' – putting life before profit. Workers responsible for the labour of social care think differently about 'value' and 'security' – this is why women must take leadership in Japan – for these things are in no way resolved.

Postscript

Since the Fukushima accident, the Tokyo Electric Power Company attempted to manage the site with an ad hoc holding operation by circulating 125 tonnes of contaminated cooling tank water each day. But TEPCO admitted its failure 'to cleanse' the Fukushima cooling water of radioactive elements – iodine, ruthenium, rhodium, antimony, tellurium, cobalt and strontium.[40] As a result, the Japanese government agreed to the release in 2023 of more than a million tonnes of this radioactive water from the plant into the Pacific Ocean. While the International Atomic Energy Agency approved the move, it outraged people in neighbouring countries, global environmentalists and especially the region's fishing communities. The decision is a major threat to Pacific Island states, original signatories of the 1985 Rarotonga Treaty for a South Pacific Nuclear Free Zone. At the same time, in response to the climate crisis, some among the international business community and a few scientists are actually promoting nuclear power as a 'clean solution' for global energy needs. In fact Al Gore came forward with such a proposition at the COP 28 Climate Summit in Dubai. The Japanese government certainly has plans for more reactors. But nuclear plants are slow to construct and expensive to run. In resource terms: uranium supplies are finite; mining dislodges indigenous communities; it contaminates groundwater; installations use massive amounts of cooling water; and waste storage is problematic. As atmospheric weather events become more unpredictable, the probability of a nuclear accident caused by lightning strikes, typhoons, wild fires, landslides, floods or rising seas increases. Japan itself is particularly prone to earthquakes. Civilian use of nuclear technology is interlinked with the global proliferation of military weapons, not to mention the vulnerability of installations themselves as targets of enemy attack. The Stockholm International Peace Research Institute (SIPRI) reported that in 2020 some 13,400 nuclear weapons were in place across the world.[41] Early in 2021, a breakthrough came when the International Campaign to Abolish Nuclear Weapons (ICAN) achieved ratification of a Treaty on the Prohibition of Nuclear Weapons. As a result of mass support from smaller nations, these are now illegal by international law. However, the Treaty signatories do not include Big Powers, all major arms manufacturers and members of the so-called United Nations Security Council. But peace activists are hopeful that ratification will happen as financial institutions begin to disinvest from the nuclear industry.[42] Yet in 2021, more bad news, as a trade deal was struck between Australia, the United Kingdom and the United States (AUKUS) for manufacture of nuclear submarines in the city of Adelaide. Moreover, this Parliamentary Bill will override the reach of Australia's nuclear protection agency known as ARPANSA.[43] Further, the desert region of Woomera will be used as a nuclear waste disposal site. The stated rationale of AUKUS is to secure South East Asia and the Pacific region in the event of Chinese territorial expansion; there are few winners in this scenario. The nuclear industry extracts a social debt from its exposed workers; a livelihood debt from colonized peoples whose habitat is sacrificed to mining and testing; an embodied debt from mothers of children deformed by radiation exposure; and a generational debt from youth as they learn how to survive in the ecomodernist *Androcene*.[44]

5

Earth System Governance

Uncertainty Principle Revisited

In the twenty-first century, the rhetorics of neoliberal freedom from big government and an open-source, boundaryless world are used to flag 'progress'. Talk about 'governance' rather than 'government' similarly reflects a preoccupation with horizontal communications between actor networks. Some Left sociologists too are 'wired' – from Manuel Castells to Michael Hardt and Antonio Negri – even replacing the proletariat with postmaterialist information workers.[1] Bruno Latour imagines a positivist world of objects and assemblages arranged across a one-dimensional plane where everything becomes pretty much equal to everything else; in contemporary posthumanism the distinction between science and ideological scientism is poorly understood.[2] Environmentalism is increasingly technocratic but at the same time rests on an ungrounded idealism in the philosophical sense. The paradox is likely inherited from the systems modelling of the Club of Rome, but by the turn of the millennium, a cyber generation in the Global North and among educated classes in the South would take for granted the use of 1/0 computer simulation 'to explain' ecological processes.[3] Many mainstream policymakers assume that the interaction of 'social systems' and 'natural systems' can be readily monitored and managed in this way. But the reliance on information theory is too simplistic to account for complex biophysical feedback cycles over time – let alone human behaviours. What follows charts a new area of political ecology, amplifying an elective affinity between patriarchal-colonial-capitalism and digitized reasoning.

Conversely, a grounded reading of this technocratic discourse shows just how parochial it is in a world where only 10 per cent of people have a car. The global structure of class, ethnic and sex/gender inequality is neatly exposed by Ulrich Brand and Markus Wissen.[4] The latter use statistics from the Bundestag to describe the 'imperial mode of living' in an industrialized country like Germany – where five kilos of resource are removed from someone else's community for every one kilo of home consumption. Beyond this, reliance on computerization to manage every conceivable aspect of daily life actually multiplies environmental damage to an extreme degree.[5] This material dependence of the Global North on Global South is known as extractivism, and it is how the so-called developed world keeps its lead.[6] Extractivism, with attendant ecological and humanly embodied debts, is also the basis of so-called 'sustainable development' models like the Green New Deal and UNEP's Green Economy proposition favoured by transnational business.[7] This profit-oriented industrial provisioning in the name of progress is materially incompatible with global ecological health, democratic futures and indeed, cultural existence.[8]

Critical political ecologists, scholars in a radical research field blending insights from geography, anthropology, sociology and ethics, will want to interrogate this new form of technocratic management.[9] A good place to start is the Earth System Governance (ESG) paradigm being networked internationally by academics from Europe. Advocates of ESG claim to present the social sciences with a major challenge: 'there is hardly any coherent, systematic, structured system of global environmental governance . . . [but] a complex web of multiple and interacting actors, networks, and institutions . . . the number and type of actors in global environmental governance has multiplied in the last decades.'[10] This governance agenda involves a postmodern proliferation of authority forms – public and private actors and new vertical and horizontal links between transnational administrative bodies – dispersing accountability while questioning the role of the nation state. ESG would fill a perceived research and management gap – but its implications for citizenship, social justice and cultural autonomy warrant scrutiny. The argument that follows contests the epistemological adequacy of ideas like ESG for governing the interplay of social and natural cycles. It highlights the difficulty of conceptualizing the humanity-nature interface for practitioners relying on algebraic systems theory and other idealist methodologies. As an alternative, it invites political ecologists to situate their work in embodied materialist approaches to the humanity-nature metabolism.[11]

ESG has a clear trajectory starting with 1970s US foreign policy recommendations from George Kennan for a global management plan located outside the United Nations.[12] International concern over societal impacts on the ecosystem was affirmed with the 1972 UN Conference on the Human Environment in Stockholm and establishment of the UN Environment Programme (UNEP). A decade later, the Brundtland-chaired World Commission on Environment and Development produced *Our Common Future* (1987).[13] The 1980s and 1990s were a time of consolidating liberal market ideology worldwide. A new conservative think tank, the European Management Forum became the World Economic Forum in 1987 with annual meetings of global leaders in Davos, Switzerland. A World Business Council for Sustainable Development largely designed the 1992 Rio Earth Summit with its Agenda 21, Biodiversity and Climate Change Conventions – all measures emphasizing voluntary corporate responsibility.[14] The creation of a Global Environment Facility within the World Bank added to this momentum for international environmental governance under capitalism. The abiding effect of such geopolitical and geoeconomic expansionism was posited a century ago in 1913 when Rosa Luxemburg observed how capital accumulation in industrial economies can only be maintained through the material subsumption of resources, labour and markets at the geographic periphery.[15] The result is that othered people's livelihood resources, traditions and conviviality are sacrificed. Nevertheless, there is fierce grassroots resistance to this global system of accumulation. Plans to institutionalize market values at the heart of world political decision-making provoked major street protests outside the 1999 World Trade Organization (WTO) meeting in Seattle.[16] At this time, the ambitious Multilateral Agreement on Investment that would have consolidated the reach of free trade was rolled back by popular dissent. Political movements from *Via Campesina* to Blockupy continue to oppose the globalization of neoliberal exploitation. Resource wars and land grabs, job losses and precariousness are provoking public unrest on every continent.

Conceptual Fit

The ESG research programme emerged out of the call for a World Environment and Development Organization made in 1998 by Frank Biermann – an international lawyer based first at the Potsdam Institute then later at the Free University Amsterdam.[17] From here, an extended academic network would grow, initially as the Global Governance Project (GGP), then transmuting into ESG from 2009. The international reach of ESG – sponsoring centres, publications and conferences around the globe – extends through universities from Amsterdam to Tokyo to Boulder and beyond. There are interest groups in many European universities with a coordinating office in Lund, Sweden. The full funding base of ESG is not transparent, but Biermann and Pattberg acknowledge the GGP as having received support from the Volkswagen Foundation, King of Spain, European Cost Action and European Union via the Potsdam Institute and Free University Amsterdam.[18] ESG spans the terrain of economy and ecology, purporting to offer opportunities for innovative problem-solving and consensus building. However, the formalization of such discursive activities is also hegemonic, holding capitalist power relations in place.[19] This context makes it difficult for ESG to deal with questions posed by the victims of free-market economics.

How does ESG create a conceptual 'fit' between social and natural systems – the better to then govern them? Oran Young, the political scientist who edits an ESG series for MIT Press, sees Earth governance as relatively straightforward: 'The essential step is to reach agreement on an appropriate structure of rights, rules, and decision-making procedures. Once that is done, it becomes timely to consider the kind of organizations needed to administer these institutional arrangements.'[20] Daniel Bromley, a thoughtful economist disagrees:

Young's insights, intuitively obvious on their face, entail the following presumptions: (1) the ideal institutional design (a management regime) must fit the problem; (2) behavioral mechanisms must pay attention to incentives; and (3) the necessary institutions (rules) must be embedded in (fit) the proper organizational structure.[21]

This question of conceptual fit is a profound epistemological challenge for academic researchers, but it is also a matter of ethics – since if policy fits material relations poorly, there will be all kinds of social and ecological fallout. Indeed, rational management of the Anthropocene under capitalism may be an oxymoron. Studies of global corporations show that they operate with 'multiple identities'.[22] Thus, corporate leaders play the jobs versus environment card, adopt 'green production' simply because it is profitable to do so, deploy new technologies only under WTO pressure or to meet mandatory Multilateral Environmental Agreements (MEAs) or they simply seek to pacify a critical media. Is this functional duplicity on the part of capitalist actors unavoidable? Can a well-intentioned scheme for Earth governance cohere, given the multiple identities of private actors – or for that matter, conflicting interests between different fractions of capital?

The optimistic ESG programme has described its central Science and Implementation Plan as organized under '5-As' each functioning at a different scale.[23] To paraphrase, these are:

- *Adaptation* – political capacity for flexible responses to new knowledge or to Earth system disturbances critical to governance.
- *Agency* – drivers and actors like businesses and NGOs.
- *Architecture* – integrating overarching governance institutions – local, regional, national and international based around shared principles for stakeholder decisions at all levels.
- *Accountability* – new designs for legitimacy reflecting an acknowledgement that financial requirements for participation in global governance may create unequal advantage.

- *Allocation and Access* – principles of justice, support and compensation as well as analysis of socio-ecological adaptability and resilience at the global level.

This framework points to the need to rationalize the 800 or so MEAs in existence – perhaps as Biermann suggests via treaty 'clustering' – under rubrics like geographic location, environmental problem, human cause, type of policy instrument or need for capacity building – each cluster requiring specific administrative interventions.[24] Biermann's priority, however, is an upgrading of UNEP's role.

What the multi-scalar architecture of ESG passes over is the living materiality and subjectivity of class-based, ethnic or sex/gendered voices, not to mention the complex metabolism of interspecies relations. The governance text deals only with institutional objectives. ESG models tend to abstract, conflate or reify grounded ecological, social and embodied processes. The ecomodern stance is that whereas the social sciences have relied on substantive historical observation, ESG will be future oriented using long data runs, statistical modelling and scenario research – involving new criteria of evidence, validity and reliability. The element of techno-scientific utopianism is confirmed by a symptomatic silence on the fact that both the internationalization of interlocked political regimes and the scientific monitoring of data will be digitalized and coordinated from outer space by GIS. But this high-tech governance is not cheap environmentally – it involves heavy metal extraction, massive energy use, toxic plastics manufacture, non-biodegradable waste and carbon-generating free trade. All these costs are rarely factored in by development proponents. Neither is the materiality of embodied debts acknowledged – family dislocations, polluted water or electromagnetically induced cancers.

As the Frankfurt School Critical Marxists – Max Horkheimer and Theodor Adorno – explained decades ago, the capitalist drive to control speaks the foundational European Enlightenment narrative of Mastery over Nature through instrumental reason.[25] This is exemplified when Biermann writes: 'No longer is the human species a spectator that merely needs to adapt to the natural environment. Humanity itself has become a powerful agent of Earth system evolution.'[26] The ESG *Androcene* is embedded in this ideology of modernization, with its notion that the step from human cause to natural effect is linear and predictable. Unfortunately, the obsolete model of classical physics, foundational to science and engineering, has been adopted by economics and even organizational sociology. The result is not science but ideological scientism and, as feminist scholars explain, the application of mechanics to partially understood living processes risks more than a little methodological forcing.[27] Nonetheless, in a neoliberal economic system geared to material accumulation, the role of technocrat professionals is to objectify, 'design and control' living human and external nature as a resource base for entrepreneurs. Noting the parallel in how mainstream modelling of economics and ecology each rely on digital parameters, several prominent scientists recently attempted to dislodge this hubris with a consciousness-raising question: 'to what extent can mechanisms that enhance stability against inevitable minor fluctuations, in inflation, interest rates or share prices . . . perversely predispose towards full scale collapse?'[28] Bromley displays a similar caution about environmental management:

> institutions (rules) and governance structures, intended to address a particular ecological problem, *necessarily set[s] in motion a new ecological trajectory whose salient properties are unknown until it is too late* to craft new appropriate and incentive-compatible institutional remedies . . . We may think of this problem as a variation of the Heisenberg Uncertainty Principle.[29]

EcoModernism

If the uncertainty principle does indeed belong here, then Earth governance on the scale imagined by ESG will call for some very skilful public relations. One might draw a parallel with Brian Wynne's work on the politics of nuclear power and biotechnology.[30] In his experience, expert management institutions tend to bolster their authority with communications that reduce social and ecological complexity. In addition, policy shaped by economic interests quantifies risk in ad hoc ways to make it tradeable against perceived benefit. The result is that citizens lose trust in governments, agencies and corporations. Even Ulrich Beck's 'risk society' thesis disguises the fact that environmental impacts are not the same across classes, ethnicities and sex/genders.[31] ESG operates with a similar 'flat Earth' model, ignoring the force of power relations in both the social construction of science and in the social distribution of its material effects. The risk society thesis is related to the discourse of ecological modernization – a European version of the optimistic post-war American evolutionist sociology of Talcott Parsons. Policy modernists and postmodernists envisage that sophisticated digital technologies will enable more efficient resource use, so decoupling economic growth from ecosystem damage.[32] This 'dematerialization' is about fine-tuning forces of production while leaving unjust and dysfunctional social relations of production intact. The US 'natural capital' school of Paul Hawken shares this engineering approach, offered as an attractive alternative to the unpalatable 'limits to growth' position.[33] However, a 2003 survey by Richard York and Eugene Rosa discovered multiple methodological and empirical inconsistencies in research studies portending to demonstrate dematerialization. Meanwhile, John Bellamy Foster highlights the implicit imperialism and unequal exchange that ecological modernizing practices rest on.[34] It is not hard to see why ecological modernization has become the dominant intellectual currency of business and liberal policy circles. By extension, the UN practice of 'mainstreaming' in development programnes homogenizes communities according to the Transatlantic consumer norm. In addressing the Earth's biogeochemical systems, ESG proponents acknowledge scientific complexity, but the economic embeddedness of their paradigm in capitalist property law demands scientific simplification. The same dilemma afflicts many areas of environmental management. In the scientific governance of genetic engineering, for example, the unitary, measurable 'gene' becomes a category for facilitating property rights, despite overwhelming evidence of the fluid interaction of non-patentable open-ended epigenetic processes in the unfolding of life.[35] Still, Biermann remains convinced of the potential for Earth governance. In his view, neither lifestyle modification, current technologies, state-directed command and control nor market mechanisms are sufficient for the necessary social transition to sustainability. ESG is judged indispensable in providing human society with a safe transition to co-evolution on a planetary scale. The social and ethical dimensions of that transition are not spelled out but reference is made to the 1987 World Commission on Environment and Development (WCED). Yet this Brundtland notion of 'sustainable development' was internally contradictory from the outset, since one part of it implied stability, while the other implied growth.[36] ESG plans to rectify this tension, by 'balancing' what the 2002 Johannesburg World Summit named the three pillars of sustainable development: Environment, Economy and Society.

ESG is intended to dovetail with the influential 2009 work of ecologists Johann Rockström et al. on 'planetary boundaries'.[37] The expectation is that inductively derived scientific data on climate change, biodiversity loss, nitrogen and phosphorous cycles, stratospheric ozone depletion, ocean acidification, freshwater and land use, atmospheric aerosols and chemical pollutants will result in benchmarks or 'threshold parameters' for political decision-makers. The German Advisory Council on Global Change adopts a similar approach to ecosystem risk – and the physical terminology is telling: 'tolerable windows', 'guard rails', 'critical values' and 'tipping points'.[38] The ESG focus on planetary boundaries is to provide a 'target corridor' for sustainable development designed to shape future international negotiations. Moreover, if boundary measures remain empirically imprecise, this should not be seen as a major conceptual shortcoming, according to Biermann, since consensus among scientists and policy networks is ultimately what matters.[39]

Planetary boundaries are understood here as 'value neutral' in themselves thereby not determining any limits to growth as in the 1972 systems approach of Meadows et al. But rendering boundaries operational in policy is clearly political, since risk assessment depends on normative assumptions. Thus in an unequal world: 'Richer societies might prefer a risk-averse approach, conserving the world as it is, and preventing any harm. Poorer societies on their part, might be more risk-taking, prioritizing economic development to alleviate poverty.'[40] Social justice activists may counter that ethically speaking, those who pollute most in the affluent North should do more to alleviate poverty in the geographic periphery of capitalism where farmers commit suicide and children starve as livelihood resources are appropriated for cash-based development. These raw complexities are not catered for in ESG, where international agreements like the 1971 Convention on International Trade in Endangered Species (CITES) or private initiatives like the Forest Stewardship Council are held up as prototypical mechanisms for Earth governance.

Having stressed the importance of planetary boundaries in defining 'the target corridor', Biermann seems to turn around, arguing that 'political institutions should follow social activities, not necessarily planetary boundaries . . . [In science] the complexities of multi-causality and multiple consequences require indeed an integrated, interdisciplinary assessment'.[41] Examples given are the Inter-Governmental Panel on Climate Change and the Inter-Governmental Platform on Biodiversity and Ecosystem Services. Interdisciplinary thinking is certainly needed, but it is only useful when sociopolitical causes and effects within a 'social system' are made explicit. At this point, the political ecologist might ask: well what exactly is a 'social system'? The ESG literature is seriously lacking in such sociological depth. Biermann believes that state sovereignty has to give way for Earth governance to be effective globally, just as it already has for economic governance via the World Trade Organization. As he sees it: 'The success of the world trade regime, for example, is related to the simplicity of its commitments, including its quantitative targets for the reduction of custom obligations.'[42] However, this is not a universally accepted line among ESG partners, particularly those from the Global South.

Reducing Complexity

What other issues and challenges are at stake in ESG? In 2012, a special ESG issue of *Ecological Economics* explored the fit between governance and biosphere dynamics – assessing international regime theory, network approaches, institutional and policy analysis, polycentric economics and resilience thinking. According to the editors' introduction, the challenge is to determine 'how governance is adaptive to the complex behavior of linked social-ecological systems at the same time as providing space for bottom-up ecosystem stewardship processes that link communities, practitioners, and scientists'.[43] An article by Padmanabhan, a planner, and Jungcurt, a sustainability researcher, addresses the objective in this way:

we aim at reducing complexity in understanding human-biodiversity relations, making cases comparable across sites, and propose that, in order to address complexity, we need a method of abstraction that leads to the development of a more structured analysis, based on selection of explanatory factors according to conceptual models as well as empirical significance. . . . [The authors add that institutions are] 'the rules of the game' . . . at the interface between the natural environment and the actors putting it into use for different ends.⁴⁴

This stylized 'use perspective' scientizes and neutralizes market activity as quantifiable 'choice sets' for firms or households. One actor's interest in use of a resource may thus cancel the option of another. Ecosystem services are described as occupying different ecological scales, with other levels of analysis identified for transactions like decision-making and appropriation. In the unique language of the authors, there is a need to focus on 'the operational, collective and constitutional choices taken at each of the jurisdictional levels of human choice'.⁴⁵

While the research would govern biodiversity by mapping 'sequential chains of interest directed transactions', the transaction idea is less about an exchange of commodities than about liberties and property rights. Take 'the gene': 'prior to its revelation through research activities, this value is highly uncertain. As soon as it has been decoded and its value identified, it becomes an information resource that has the characteristics of a pure public good.'⁴⁶ Thus the traditional farmer's use perspective contrasts with commercial breeder's interests, but reconciliation is envisaged through skilful governance. Given this functionalist framing, the article does not discuss power differentials under capitalism and the prevalence of corporate biopiracy. In fact, the research illustrates the postmaterialist vacuum that occurs when algebraic models are used to answer social questions. Ironically, as indicated in the quote, the 'empirical deficit' is to be resolved by further abstraction. It is therefore curious that the authors endorse and paraphrase Anil Agrawal to the effect that 'Theories of collective action phenomena in biodiversity governance have little explanatory power beyond the specific empirical setting in which they were conducted, as relevant causal models are missing'.⁴⁷ In relation to ESG objectives, it is also odd that Elinor

Ostrom is referenced – a thinker renowned for arguing that local neighbourhoods manage the commons far more effectively than organizations or markets.[48]

The game theoretic scientism marking ESG analysis is entirely dependent on the manipulation of idealized units, so raising the problem of 'conceptual fit' again. On this epistemological dilemma, the authors quote a further opinion that

> Biophysical attributes do influence transaction properties by virtue of their material resource characteristics, but, as explained above, they are not the same as the cognitive conception of the properties of transaction. Properties of transactions [as seen in the ecosystem include] . . . jointness and absence of separability, coherence and complexity, limited standardisability and calculability, dimensions of time and scale, predictability and irreversibility, spatial characteristics and mobility, adaptability and observability, etc.[49]

Padmanabhan and Jungkurt seem unresolved as to which phenomena are ontological existents 'observed in the ecosystem' and which are socially constructed human abstractions shaped by the language of physics and projected by the investigator on to living material. This cognitive flaw typifies the post-Enlightenment shift from sensuous human interaction with natural processes to atomized measurement for purposes of calculated control. The global environmental crisis experienced today testifies to the error of this instrumentalism as a false historical consciousness. And indeed, journal editors Galaz et al., seem to affirm this conclusion, writing to the effect that 'management based on thresholds, although attractive in its simplicity, allows pernicious, slow and diffuse degradation to persist nearly indefinitely'.[50] The governance of metabolic flows by disentangling them contrasts sharply with the embodied materialist approaches of indigenous farmers or caregiving mothers who protect 'natural and human' complexity by labour forms that enhance cycles of reciprocity.[51]

Steering Laissez-faire!

On the question: Can Earth systems be governed? Nilsson and Persson – supported by the Stockholm Resilience Centre, Future Foundation and Mistral

– offer a more nuanced position. They propose 'policy packages' to encourage economic transformation and reduce system stress, risks and vulnerabilities. They also avoid simplification, acknowledging that natural processes are synergistic, such that human interventions may well result in shifting the material burden from one ecosystem process to another or one country to another. They recognize a variety of potential governance modes: regulatory standard setting and monitoring; market-based economic incentives; normative or consensual practices, but maintain that market pricing of 'ecosystem services' can skew economic signals and lead to problem shifting. They also concede that research tends to be framed by the assumptions of socially dominant groups. Finally, Nilsson and Persson believe that regulation is necessary but accept Ostrom's view that policy fails if too centralized. That said, satellite monitoring is recommended 'to complement' local knowledge. After a steady to and fro on the vicissitudes of Earth governance, these authors arrive at a rejection of ESG logic altogether:

> biophysical interactions are of such complexity that they cannot possibly be orchestrated in a synoptic way . . . The interactions that occur within and across natural, social and economic systems and shape *the unfolding pattern of global environmental change can neither be predicted and controlled top-down*, nor left to market dynamics under some generic regulation.[52]

While the ESG model is oriented to an analysis of self-steering structurally autonomous systems, the world of its actors is competitive and entrepreneurial.[53] In this schizoid methodology, 'systems' must be predictable to be manageable, but the behaviour of the very individuals who cause these systems to come into being is free and unpredictable. Sociologist Timothy Luke, extrapolating from Michel Foucault's thesis on biopower, has commented: 'Most sustainable development discourses are extremely conflicted and as discourses of green governmentality, they are often little more than a bureaucratic conceit, indulging the empire-building of professional would-be eco-crats.'[54] These epistemological flaws are not likely to unsettle the winners of neoliberal globalization – bankers, CEOs, hedge fund operators or technocrat professionals. But others from the global and domestic peripheries may be very uneasy about forms of scientific mystification – or scientism – being used to promote the values of free trade and unequal North/South exchange.

The 'overarching' ESG management regime for mitigating the ecological crisis implies that nature can be measured and valued relatively easily. The approach follows neoclassical economics where value is founded on individual self-interest. Thus, the value of something is 'revealed' in the rational choice of what an individual or group of individuals are willing to pay for it. Under this model, environmental resources are goods, stock or 'natural capital' that can be measured along a single scale, thereafter translatable into money. Just as markets are said to be driven by individual actors behaving as units, so the Earth is believed to be made up of functioning units or things commensurable by dint of the capitalist exchange principle. A thing that is not commodified has no value – it is 'an externality' to the economic system. Problems are believed to occur when an aspect of nature has no market price. But once commodified, as in water pricing or pollution taxes, 'a thing will be managed spontaneously by "the hidden hand" of the market'. Creation and enforcement of property rights becomes the solution to resource management. Now critics have long contested this viewpoint, recognizing the utilitarian calculus as an expression of capitalist class interest. At a deeper level, as ecofeminists point out, the discourse of exchange value is sociologically sex/gendered 'masculine'.[55] Peoples from the Global South express their pride in conservation of habitat through customary law.[56]

Value is far more complex than implied by the ethic of methodological individualism, where even altruism is an algorithm of utility and self-interest. In environmental economics and sometimes in ecological economics as used by business, government and policy academics, the exercise of valuation is very ideological. In addition to believing abstract Nature is unitary, fitting an ordinal scale, the approach often equates the natural material world with the humanly known world; essentializes all humans as self-maximizers; treats the symbolic and cultural realm as irrelevant to value; prioritizes minor human interests over major non-human ones; reifies and then manipulates economic weightings; assumes that humans can substitute or offset natural capital; defines sustainability in a non-ecological way; and attempts to wield authority by constructing 'consensus'.[57] Then there is the rather bogus choice between 'weak sustainability' – preserving capital – and 'strong sustainability' – preserving 'natural capital' – in the environment out there, although still under capitalism. The conventional tools of Environmental Valuation are also functionalist closed-system activities: Cost-benefit Analysis and Contingent Valuation are marred by reductionism; Multi-Criteria Mapping and Multi-Attribute Utility Analysis neutralize judgement; Rational Dialogue and Deliberative Democracy understate the force of power relations in the social construction of consensus.

The contemporary context of these power relations is financial and ecological chaos wherein the ambition of Earth System Governance can too readily bend to a transnational ruling class seeking to stabilize conditions for accumulation through global integration. Political ecologists may reject the very idea of ESG as *zweckrationalität* or instrumental rationality on a planetary scale; but insiders see it as open, self-organizing, non-hierarchical – inclusive of inputs from business, indigenous NGOs or other 'non-state actors'. There is convergence with the neoliberal small state ideal and a 'science-policy interface' built on PPPs – public-private partnerships. ESG foci include UN reform, particularly 'treaty congestion', and support for current international 'green' designs like the World Bank-UNEP 'bio-economy' which rests on the commodification of ecosystem services, neocolonial technology transfers and even risk-based climate derivatives. ESG research can readily serve to legitimate or ideologically naturalize top-down political control through academic credentialing.

Validity

In order to defend its environmental reputation, the corporate sector relies on a notion of 'sound science' in product marketing and project approvals – and as Kees Van der Pijl has pointed out, technocratic professionals command good salaries and play a key role in manufacturing that validation.[58] However, many argue that whereas business interests 'technicize' arguments, what is wanted is open discussion and sharing with those who live with the impacts of extractivism, industrial development and free trade.[59] Certainly, a shift is occurring towards more inclusive sustainability deliberations where contextual knowledge and subjective inputs are valued. Following the model of Silvio Funtowicz and Jerry Ravetz, the post-normal scientist or policymaker recognizes that problem definition, choice of what gets measured and decisions on commensurability have a normative and political aspect. He or she will reach beyond consensus within a specialist 'epistemic community' to learn from the experiential skills of othered groupings – peasants, for example, even grandmothers.[60] But can such deliberations ever be effective for Earth governance if contained by neoliberal institutions?

The climate change debate shows that democratic policy is easier to claim than to achieve. Initially, the Inter-Governmental Panel on Climate Change (IPCC) lacked grassroots participation and cross-cultural sensitivity, thereby neglecting Global South perspectives. Like most multilateral governance agencies, the IPCC accepts 'the imperial mode of living' as human norm. Hence the emphasis on climate change 'adaptation' for developing countries – as distinct from a materially effective remedy like recommending rapid degrowth for the affluent North. When the IPCC does activate public validity criteria, it invites governmental input, with the outcome that its synthesis reports hover somewhere between science and diplomacy, as Sheila Jasanoff observed.[61] The ESG programme is similarly hamstrung by subservience to capital. What is taken-for-granted is precisely what needs to be challenged. The ecosocialist analysis exposes the root of the matter by demonstrating how capitalist industrial production and the rise of cities led to a major metabolic rift between human economies and their environmental base.[62] By this thesis, capitalist commodification of nature and the treadmill of production-consumption are inherently incompatible with sustainability. Further, the commodity society generates a one-dimensional monoculture of the mind. Carbon trading is but the latest in a long line of efforts that simply reinforce the prevailing crisis while affording new opportunities for 'accumulation by dispossession'.

Finally, the hegemony of digitalized reasoning comes to colonize the environmental movement North and South. As Tim Luke comments, the subsumption of culture and ecology under economics is now found even among grassroots NGOs, when they posit that 'existing flows of energy, information and resources are not wrong as such, only that the business and administrative elites of nation states are mismanaging their volumes, rates, and levels'.[63] It does not have to be this way, of course. There is an alternative to the short-sighted notion of Earth governance that is the focus of this chapter. Articulate voices speak out from the geographic, and indeed domestic, peripheries of capitalism, many of them inspired by the Zapatista model of democratic eco-sufficiency. They are joined by social scientists like Manfred Max-Neef or Veronika Bennholdt-Thomsen.[64] The critical fact is how ecologically rational the labour of peasants, indigenous foraging peoples and household caregivers is, as they provision without the exploitation and entropy that results from large-scale profit-directed production. Here is an embodied materialist epistemology, an empirical science and a living model ecological economics that shames the androcentric sensibility of ESG.

To recapitulate: the academic weaknesses highlighted in this critique of the ESG mission have two facets. One set of problems stems from the historical context of a research programme that is embedded in the political norms of neoliberal globalization. A second set of problems stems from reliance on post-Enlightenment knowledge disciplines whose epistemological capacity for dealing with Nature is compromised by the master narrative of instrumental reason. These two problem sets are mutually reinforcing – which is no surprise given the elective affinity of patriarchal-colonial-capitalist norms. Plainly, ESG researchers are unfamiliar with or choose to overlook a vast social scientific library and contemporary platforms like the *Journal of Political Ecology, Environmental Ethics, Globalizations, Capitalism Nature Socialism, Journal of World-System Research* or *Antipode*, to name a few. A political ecology informed by the critical Marxist tradition and grounded varieties of poststructuralist thought has longstanding commitments to local-level research, habitat protection and social justice. Political ecologists are well placed to educate thinking publics, academic scientists, policymakers and activists on the implications of initiatives like ESG. The Earth System Governance project has fired the imagination of scholars even as it has underscored the dangers of academic work that neglects its own context. An easy interdisciplinarity based on the algebraic scientism of systems theory will not expose the global power relations that underpin social and ecological crises today. The same idealist abstraction occurs in talk about natural capital and dematerialization, and it reflects a deep dissociation from the sensuous life world. For the imputation of economic exchange value to Earthly metabolisms only translates thermodynamic flows into fictional stochastic units leaving material ecologies behind. Computer-dependent techno-utopian governance can only perpetuate the destructive biogeochemical circuits of industrialization. As such, environmental and health costs accrue as debts to less powerful humans, unborn generations and the Earth itself.

6

The Gene Trade
Organized Irresponsibility

The denial of human embodiment in living ecologies is foundational to patriarchal thinking, including modern economics and science, ethics and law. In the practice of genetic engineering, this dis-valuation of material nature – what can be called the *1/0 imaginary* – is revealed at multiple levels:

- the ontological split of Humanity/Nature
- the epistemological schism between determinism and complexity genomics
- the methodological fracture between fact and value in risk assessment
- the legal disconnect between patent originality and substantial equivalence
- the policy elision between coexistence and colonization
- the ethical contradiction of us versus them.

The androcentric ideology of genetic science is very apparent in Erwin Schrodinger's assertion that chromosome structures are 'code, law and executive power, or to use another simile, they are the architect's plan and the builder's craft in one'. For genetic researcher Craig Venter, a holder of multiple GMO licenses in conjunction with Exxon Corporation and the United States Department of Defense, 'life is a DNA software system'.[1] By contrast, given this Promethean certitude, cell biologist Stuart Newman emphasizes that gene interactions are actually quite non-linear and often epigenetic. 'Most biological features are the collaborative products of many genes ("multifactorial") that work in conjunction with non-genetic factors.'[2]

It is no surprise that the late twentieth century saw a new generation of patent or perish scientists, along with the replacement of independent science media by a public relations industry.[3] The Australian government, encouraged by the Department of Foreign Affairs and Trade, has been drawn into the economics of biotechnology since the early 1990s. The major political parties, Liberal and Labor alike, are growth oriented, pro free trade and pro the development of genetically engineered products. At the turn of the millennium, an Office of the Gene Technology Regulator (OGTR) was established under the Commonwealth Department of Health and Aging, although other departments like Agriculture or Environment and Heritage could make input to its agenda.[4] Given that human well-being depends on ecological conditions, the administrative arrangement likely reflected the international power of pharmaceutical corporations, as Big Pharma increasingly defines the political agenda of nation states.

Measurable Units

In Australia, a *Gene Technology Act 2000* was established to protect the health and safety of the nation's people and environment by recognizing risks posed by genetically modified organisms (GMOs) and then managing these new life forms by regulation. The OGTR grants accreditation to research institutes, certifying private and public facilities to invent 'novel organisms' and to license commercial dealings. Hundreds of biomedical research projects have passed through the OGTR's office, but scientific uncertainties in genetics and difficulties in carrying out risk assessment raise tough questions. Already by 2005, the Regulator had approved hybrid crops for cultivation across the country: canola, Indian mustard, grapes, clover, cotton, sugar cane, pineapple, papaya, peas, poppy. In this driest continent on Earth, a major commercial release was Roundup Ready cotton – heavily reliant on irrigation. Meanwhile, genetically engineered 'drought-resistant' wheat and a herpes vaccination for cattle were being looked at.

What needs serious consideration here is the fact that the environmental release of a novel organism may break evolved food chains in nature so leading to species extinctions – already reported at a massive 60 per cent globally. The iconic Koala bear habitat, for example, could fail as a result of genetically engineered eucalyptus trees. In the human diet, foods derived from plants engineered to coexist with herbicides like glyphosate carry excessive chemical residues. GM corn filler in processed foods can set off allergies, gut problems or unforeseen inheritable complications. North American laboratories are known to have used bananas for rabies antibody production, but how safe is such 'pharming' in a world where competitive food markets are the norm? Another risky proposal has been the breeding of pig hearts for human medical transplants, since the porcine virus can infect a recipient, and even their neighbours, with outcomes that are potentially epidemic in scale. Beyond this, experimentation with smallpox and anthrax points to the overlap of genetic engineering with R&D for biological weapons. The National Institutes of Health even reports the US Army using secretions from GM hamster kidneys and cow mammary cells in designing bulletproof vests from spider's silk.[5]

Biotech research is risky for many reasons. The DNA in engineered organisms is highly unstable and thus unpredictable, because it uses promiscuous viruses, or *E. coli* bacteria, as carriers of DNA from one cell to another or one species to another. The Australian *Gene Technology Act* requires the OGTR Regulator to be satisfied that risks to the public from GM experimentation can be managed before a license is issued. However, genetically hybrid crops are not assessed in the field. In one case, the Commonwealth Scientific and Industrial Research Organization discovered that commercial Bt cotton, injected with the *bacillus thuringiensis*, not only exudes the Bt toxin through leaves, so killing off insects as planned, but exudes it via roots as well, knocking out the bacteria colonies that keep soils viable.[6] Beyond this, the very practice of 'risk assessment' as a decision-making tool is inherently problematic. Already by 1998, the opinion of a United Kingdom Royal Commission on Environmental Pollution was that no satisfactory way exists for measuring environmental risk nor any way to decide on what scale of risk is humanly tolerable.⁷ The risk assessment of genetically engineered products is not made easier by the fact that scientists have disagreements over the nature of 'the gene' as such. The Australian *Gene Technology Act* takes genetic determinism as a given. Known in the trade as 'the central dogma', this theory builds on the famous 1953 Watson-Crick idea of 'a master gene' unchanging over time; determining one function; and not affected by its surroundings.⁸ In the alternative scientific model known as 'complexity genomics', the gene is treated as dynamic and open-ended – characterized by moving chromatin coils, interactions between redundant or poorly understood junk DNA, unpredictable jumping elements or transposons and newly discovered RNA. The genomes of both germline and somatic cells may mutate over time and respond to environmental factors.⁹

Unpredictable Risk

The *Regulations on Dealings with GMOs* fail to engage with this complexity, although many scientists warn that if vector viruses or bacteria stray, horizontal gene transfer may result in ecological hazards and public health crises of epidemic proportions. International members of the Human Genome Research establishment now concede that multiple factors in cell biochemistry must direct DNA, since there are not enough genes in a genome to account for all the proteins made in cell reproduction. These developments in the science of genetics cast a shadow on much industrial biotech research resting on 'the central dogma'. Despite the fact that the scientific community does not have a generally agreed understanding of what 'a gene' is, the reductionist notion of a fixed biological unit prevails because it is necessary for the law of corporate patenting and market economics. A related legal curiosity is the occasional assurance from industry that a novel product is 'substantially equivalent' to its natural origin. This ambiguous phrase seems to undermine the logic of patenting altogether but is used in the event of a pending compensation claim.

Very early in the day for genetic engineering, feminist academic Sheila Jasanoff described the interface between research and application as marked by many levels of scientific indeterminacy:

- horizontal gene transfer impacts on the water cycle, soils, plants, animals, humans or across all kingdoms

- artificial isolation of a species from essential but unknown metabolic linkages within a given habitat

- limited generalizability due to the size of the experimental area

- limited generalizability due to inappropriate temporal scale of trials

- inability to take account of non-quantifiable synergistic interactions between biota[10]

In the laboratory setting, the containment of living processes can fail in greenhouses with earthen floors, or caged transgenic fish in open ponds, or experimental vaccinations with transgenic viruses, and so on. The OGTR Regulator does request researchers to detail whether the viral vector used in genetic manipulation will remain in the final GM construct or be removed. But in the face of 'informed doubt', how are government advisors and policymakers expected to make 'ethical' decisions on behalf of the public?

Most scientists and ethicists agree that where uncertainty exists, science, government and citizens should adopt precaution – like the care taken by mothers towards small children or farmers towards field stock. The OGTR has put a version of the precautionary principle in the *Gene Technology Act*, but its role in the regulatory process is not spelled out. Internationally, the precautionary principle is embedded in the *Convention on Biological Diversity* and *Biosafety Protocol*. The EU and some seventy-nine nations support this *Protocol*; however, Australian and US negotiators have worked together in international meetings to dilute it, in favour of the pro-business neoliberal World Trade Organization (WTO) approach.[11] The OGTR also works in with local GM protocols under the aegis of a Federal-State Ministerial Council. However, the latter has not met very often. *Consequential Amendments to the Act* provide that most pre-existing government authorities will take advice from the Regulator on GM dealings. These are: the National Agricultural and Veterinary Chemicals Register; Food Standards Australia; the National Industrial Chemicals Notification and Assessment Scheme; the Therapeutic Goods Administration; the National Health and Medical Research Council;

the Quarantine and Inspection Service; and authorities at state level. Clearly, such a widely dispersed process will not be seamless. For example, while the assessment of DNA in a new product is handled by the OGTR, the chemical aspect falls under the Pesticides and Veterinary Medicines Authority. This means that the 'interaction' of DNA with chemical components is rarely examined prior to commercial licensing.

Matters Outstanding

Before licensing a novel crop or medicine, the Regulator is required to take advice from a Ministerial-appointed Gene Technology Technical Advisory Committee (GTTAC). Two further advisory bodies, a Community Consultative Committee (GTCCC) and an Ethics Committee (GTEC), have had no direct influence on OGTR decisions. Thus in 2003, while the Community Committee was alerting the Regulator to widespread disquiet among the population over GM canola, scientists on the Technical Committee advised the Regulator to go ahead with a crop license. Risk assessment by the OGTR does not look at all synergistic interactions between biota in the environment, nor at long-term human health impacts. This is not done, because technically, it cannot be done.[12] Essentially, risk assessment, 'best practice' as it is called, comes down to the scientists' best guess.

This is why the 1998 UK Royal Commission came to the conclusion that in a democratic society, 'When environmental standards are set . . . decisions must be informed by an understanding of people's values.'[13] The same reasoning should apply in the health field, of course, but the Australian GM regulatory process has not taken social values very seriously. The Community Committee and the Ethics Committee, always marginalized by the commercial focus of the OGTR, would soon enough be reduced and combined as one committee-GTECC. Nor does the Regulator take on board just how the safety of a novel release depends on sociological factors like public attitudes, organizational norms, political climate and even international trade. The point is that 'risk assessment' is a cultural process. Notwithstanding the positivist premise that an objective scientific practice must separate facts from values, ethics enter into a scientific practice each time it comes to selecting research techniques and technologies; defining what constitutes the normal life cycle of an organism; choosing to cull an animal; deciding which biophysical units to treat as commensurable; determining which risks matter and which don't; and in choosing to compartmentalize risk assessment and risk management.

It is possible that the Regulator's office in Canberra has been insufficiently staffed for providing more thorough assessment, monitoring, prosecution – and therefore full compliance with *The Act*. It is believed that its functions have sometimes been outsourced to Biotechnology Australia, which unfortunately is the federal government's GM promotions wing. A tacit 'us versus them' culture in bureaucracy does not make for transparent governance; and the public should be informed when responsibilities are being delegated across departments. People have a right to know who the backroom consultants are. These concerns have not been resolved in reviews of the *Gene Technology Act 2000*. Additional concerns raised by observers of the GM industry highlight:

- patent controls
- the legal liability of GM producers
- the principle of food sovereignty
- organic farmer's rights
- corporate use of indigenous knowledge
- the welfare of pregnant women.

Coexistence

In 2003, as negotiations for an AU/US Free Trade Agreement were going on, an ad hoc Gene Technology Standing Committee (GTSC) of the Federal-State Ministerial Council overseeing GM called for a policy principle on the coexistence of GM and non-GM cultivation on adjacent lands. The public consultation brief on coexistence did not canvass impacts on farmers; nor the extent of consumer resistance to modified foods; nor the gross economic value to Australia of organically grown crops; nor were potential debts to the public in terms of human health or environmental damage factored in. Common Law renders GM farmers liable for unintentional contamination of land – corn and canola being especially active cross-pollinators. But the victim has to establish indisputable proof of harm. The international literature raises yet further matters between the GM farmer and corporate supplier. For example, a seed purchase contract may expose the farmer to prosecution over equipment usage and storage. Corporations or governments may want to see fields and record books. Farmers may be penalized for saving patented seed, whether purchased or wind borne. From the US Mid-West to India, farmers have been losing out to this corporate colonization of their lands by unruly seeds.

Big Pharma companies like Syngenta, Aventis and significantly Monsanto, now owned by Bayer, control over 90 per cent of global GM seed cultivation. At one point, Syngenta applied to patent forty varieties of rice across 115 countries, and had this not been blocked by environmental NGOs and farmers, it could have annexed the gene sequence of twenty-three major food crops.[14] Syngenta withdrew but not before the mesh of agencies that determine what people eat was exposed. These are the European Patent Office (EPO), the World Intellectual Property Organization (WIPO), the US Patent and Trademark Office (USPTO), the Food and Agriculture Organization (FAO) and the Consultative Group on International Agricultural Research (CGIAR). In Australia, *The Gene Technology Act* contains no provision to appeal the Regulator's decisions. There is scope for a performance appraisal of the Regulator by the Australian National Audit Office (ANAO) and indeed, public submissions have pointed to anomalies. For instance, while the Regulator is required to ensure the fitness of a person to hold a GM license, companies with a record of convictions – Bayer and Monsanto, for example – have had their dealing applications approved.[15] In terms of monitoring engineered products in the living environment, the Regulator has not prosecuted any firms, although dozens of license breaches have occurred.

The stated object of the *Gene Technology Act 2000* is 'to protect the health and safety of people, and to protect the environment, by identifying risks posed by or as a result of gene technology, and by managing those risks through regulating certain dealings with GMOs'.[16] What is remarkable about this object is the admission that biotechnology is risky; equally remarkable is the unqualified assumption that risks can be identified and managed. For it is already conceded in government circles that harm resulting from premature commercialization of genetically engineered products may well be irreversible. Surely it is time to ask:

- If fundamental scientific constructs are undecided, how can risks be identified?

- And if risks cannot be identified, how can they be managed?

In this epistemological vacuum, the *Gene Technology Act 2000* required its GTEC committee to deliver an ethical formula on GM, acceptable and applicable to researchers, business leaders, scientific IBCs, policy planners and publics. *The Act* appears to seek a 'code of conduct' or ethics checklist for the Regulator. But in a globalizing patriarchal-colonial-capitalist world, Greek philosophy, Christian platitude, liberal human rights and voluntary moral competencies are not enough. A culturally and ecologically sensitive policy deliberation might have product applications reviewed by community-based citizen juries on a 'Do we need this?' basis. The biopiracy of indigenous knowledge could be examined using principles embedded in *ILO Convention 169*.[17] Philosopher Shiv Visvanathan calls this approach a 'vernacular audit'.[18] As things stand, people's submissions to public inquiries are largely ignored as governments push ahead with propositions from Big Pharma. In Australia, this includes a Synthetic Biology Future Science Platform at the Commonwealth Scientific Investigation and Research Organization (CSIRO) – more novel organisms for the 'web of life' to adapt to.

Andro Science

Public awareness of how the human gene pool is being opened up to manufacture and marketing took a leap forward with the Covid-19 pandemic. During his time as president of the UK Royal Society, Professor Robert May roundly condemned virological experimentation. And indeed, the US Centers for Disease Control and Prevention, a major holder of research patents in this area, does enlist the precautionary principle, as does the US Center for Arms Control and Non-Proliferation. Dutch legal scholar Britta van Beers observes rising investor excitement over human cloning by synthetic biology using computer-designed DNA to generate human-animal hybrids, plants and microbes.[19] In Australia, the Office of the Gene Technology Regulator OGTR is pragmatic, with fast-track self-assessment by commercial interests allowed.[20] A particular concern was the Labor government's support for the Mitochondrial Donation Law Reform Bill 2021. Ecofeminists have focused on this aspect of reproductive technology for a number of years, and as sociologist Renate Klein explains, the innovation hinges on what is known as Clustered Regularly Interspaced Short Palindromic Repeats (CRISPR).

> CRISPR is a guide molecule made of ribonucleic acid (RNA) and Cas9 is a bacterial enzyme. The CRISPR RNA is attached to Cas9 so as to work as molecular scissors. This fast new gene editing technology can make changes to early embryos that will irrevocably be passed on to the next generations.[21]

In humans, CRISPR enables the cut and paste of genome material from the good ovum of one woman into the faulty ovum of another. The original disquiet over eugenic sterilization of disabled or racialized populations has receded as middle-class consumers seek eugenic techniques to remove inheritable anomalies from their offspring. Klein points out that given the availability of prenatal tests, there is really no need for 'genetic surgery' of this kind. Nevertheless the CRISPR option is packaged and sold to women as a discounted add-on service by *in vitro* fertility clinics. In fact, with material from 'three parents' in the 'genetic donation' mix, collateral damage is a risk. In China, researcher He Jiankui was jailed in 2020 for CRISPR editing to achieve HIV resistance in the genome of twin baby girls. But when close examination of his data showed a mosaic of off-target effects, he was imprisoned by the Chinese government.

Another term for CRISPR engineering currently circulating in the scientific literature is Gain-of-Function (G-o-F), although some judge this to be a somewhat euphemistic descriptor of biological research aimed at increasing the virulence and lethality of pathogens and viruses. GoF research is government funded; its focus is on enhancing the pathogens' ability to infect different species and to increase their deadly impact as airborne pathogens and viruses. Ostensibly, GoF research is conducted for biodefense purposes.[22]

A well-known example of this is the US National Institutes of Health (NIH)-funded genetic manipulation of avian bird flu, which had the effect of enabling it to cross into mammalian species. Certainly many problems attach to G-o-F for commercial use, not least the possibility of harmful material leaking from a lab. Such laboratories are ranked by level of research security, and across the United States from New York to Seattle there are some 200 high biosafety containment facilities. However, according to the *USA Today*, not all experimentation centres are found to measure up.

> Vials of bioterror bacteria have gone missing. Lab mice infected with deadly viruses have escaped, and wild rodents have been found making nests with research waste. Cattle infected in a university's research experiments were repeatedly sent to slaughter and their meat sold for human consumption. Gear meant to protect lab workers from lethal viruses such as Ebola and bird flu has failed, repeatedly.[23]

In June 2019 the Centers for Disease Control ordered the closure of Fort Detrick, the US Army Medical Research Institute of Infectious Diseases, after finding biohazards there. Public health advocates, including Physicians for Social Responsibility, argue strongly against this direction of research citing the Nuremburg Code.

President Obama suspended G-o-F research on SARS, MERS and avian flu viruses – with the exception of Harvard, University of North Carolina and the Wuhan Institute of Virology. The Chinese laboratory worked on bat-to-human transmission funded by the National Institute of Allergy and Infectious Diseases (NIAID) under Dr Anthony Fauci's direction, with moneys from the Bill and Melinda Gates Foundation's GAVI vaccination fund. The NIH and US Agency for International Development (USAID) were also involved in the programme.[24] Following the outbreak of coronavirus in 2020, an investigation of techniques used in Wuhan was made by Dr Richard Ebright, a Rutgers-based molecular biologist and biodefense expert. By his account:

> The researchers report having conducted virus infectivity experiments where genetic material is combined from different varieties of SARS-related coronaviruses to form novel 'chimeric' versions. This formed part of their research into what mutations are needed to allow certain bat corona viruses to bind to the human ACE2 receptor.[25]

So did this experimental production of multiple chimeric forms account for the different strains of Covid-19 like Delta and other versions that unfolded as the pandemic moved across the global community? Or does the variety in Covid-19 manifestations simply reflect the instability and mutability of a man-made genome?

Certainly, the baseline virus SARS-CoV-2 exists in numerous versions, some seventy among them having been patented in the United States over the past two decades. These were documented by assets underwriter Dr David Martin, lawyer and CEO of the firm M-Cam International Innovation Risk Management. He claims to have uncovered the patent history of Covid-19 while monitoring researcher compliance with the Geneva Protocol on chemical and biological weapons. In fact, Martin appears to open a new chapter in biopolitical economy when he writes:

> Together with CDC, NIAID, WHO, academic and commercial parties (including Johnson & Johnson; Sanofi and their several coronavirus patent holding biotech companies; Moderna; Ridgeback; Gilead; Sherlock Biosciences; and, others), a powerful group of interests constituted what we would suggest are 'interlocking directorates' under US anti-trust laws.[26]

In response to the alleged accidental breach of disease matter in Wuhan, a US vaccine was released for public use in the pandemic but before full lab testing was carried out. In turn, the US Center for Disease Control's Vaccine Adverse Event Reporting System (VAERS) adopted for diagnosis and management of Covid-19 was not helped by international use of a PCR test, whose accuracy was limited by the fact that it had been developed for a different purpose according to its inventor.[27] The vaccine appeared to rely on the self-propagating capacity of viruses themselves as instruments for spreading immunity.[28] However, the effect of G-o-F intervention on the natural human immune function is not fully understood. For example: will the spike protein in the newly engineered vaccine product enter the human germline? In the event of such injury, product manufacturers of the Covid-19 vaccine took care to indemnify themselves from prosecution right at the start of the pandemic.

In spite of ongoing methodological uncertainties behind cutting-edge research in synthetic biology, professional affiliates of bodies like the UK Wellcome Trust or the US National Academies of Sciences, Engineering and Medicine accrue status and financial reward as pioneers in the field. Rather than an international treaty to regulate the technology, they plan to adopt 'a responsible pathway forward' to manage genetic engineering risks. The International Bioethics Committee of UNESCO is said to hover between a moratorium on embryo modification and an outright ban. On the other hand, the European Convention on Human Rights and the Spanish Oviedo Convention each have strong biomedical law. In fact, one Spanish company, Embryotools, relocated to Greece to avoid regulation. Methodology aside, ethical questions on germline manipulation remain, and it is a slippery slope from therapy, to eugenics, to the development of biological weapons. It is difficult to know what to make of the UK Nuffield Foundation speculation that species enhancement by CRISPR technology might offer a promising biopolitical response to the Anthropocene crisis. That is to say, future generations might be bred with a 'gained functional tolerance for adverse environmental conditions (such as those that might be envisaged as a result of climate change or in space flight)'.[29] Are they serious?

In Australia the gene trade is governed by national competition policy, the WTO and bilateral free trade agreements. However, citizen networks such as GeneEthics are working hard to hold the industry and its regulators to account. Currently, the spotlight is on OGTR approval of virus resistant bananas: a bionic product with no history of safe use. The risk is that antibiotic marker genes used in developing the product may cross with pathogens in the environment or even in the human gut. GeneEthics is also calling on

Food Standards Australia and New Zealand (FSANZ) to reject GM meat, since its growth factors may promote cancer spread. A further concern is Ultra-Processed Foods for infants using genetically engineered microbes. The Australian network is one among many global civil society action groups opposing synthetic biology; these include the Stop Designer Babies Coalition and the GMO Free Regions Alliance.

7

Buen Vivir

Ecomodernist or Andean?

Strategy I

Under the global ideology of ecomodernism, governments, international agencies and activists continue to proffer economic remedies for ecological ills. The mainstream political discourse contains no vocabulary for the metabolism of nature that reproduces human societies – and as noted this dates back to the European Scientific Revolution, which shifted the definition of nature from 'organism' to 'machine'. The mindset is clear in how this German Federal Minister for Environment outlines his vision of the future global economy as a clockwork whose parts tick over in perfect harmony:

> The linchpin of a model of sustainable development has to be a 'third industrial revolution', at the centre of which is energy and resource efficiency ... If China becomes the 'world's workbench', India casts itself as the 'global service provider', Russia develops into the 'world's filling pump', and Brazil as the 'raw materials warehouse' and 'global farmer' – provides Asia's industrial and service companies with iron ore, copper, nickel and soybeans; Germany should then assert and strengthen its position in the global division of labor as 'the responsible energy-efficient and environmental technician'.[1]

Beneath this economic machine, Nature is merely a 'raw materials warehouse'. It is therefore no surprise that global climatic patterns fail, as living ecosystems are subjected to the *Androcene* imaginary.

The minister's crude blend of neoliberal economics, technological innovation and environmental sentiment typifies the popular ideology of ecological modernization – common among bureaucrats and some establishment academics.[2] Proponents of ecomodernism, Arthur Mol and David Sonnenfeld, for example, assume that capitalism can be made sustainable.[3] But sociologists Richard York and Eugene Rosa soon pointed to the illusory features of its management constructs, which operate in a thermodynamic vacuum.[4] As governments and quasi-policy agencies like the Inter-Governmental Panel on Climate Change (IPCC) have prevaricated over global climate solutions, the transnational ruling class is reluctant to let go of its wealth, privilege and control. Climate deniers and catastrophists alike insist that 'there is no alternative' to the 1987 Brundtland Commission's environmental stewardship through economic growth and 'trickle down'.[5]

Al Gore's sustainable America plan was also full on economism but bankrupt ecologically. He envisaged congressional incentives to support solar, wind and geothermal spots in the south-western deserts of the United States. A national low-loss underground grid would be built; there would be plug-in hybrid cars, retrofitted buildings and household conservation advice. Gore would replace the Kyoto Protocol with a treaty that caps carbon emissions ready for trading. Looking at the real bottom line, the construction of new high-tech cities in the US South-West would consume vast amounts of front-end fuels – in mining metals, welding turbines and grids, road making, water supply, component manufacture for housing, air conditioning for supermarkets and schools.[6] Here you see a mortgage – borrow now, pay later – an ecological debt whereby another biodiverse ecosystem will be emptied out. In this false green conversion, poor and marginalized communities from the populous US East Coast would have to weather the psychological costs of mass resettlement, while the new solar urbanization means a loss of US food-growing land, possibly to be replaced by US agricultural leases in Central America. How then would the newly landless subsistence farmers of Mexico or Costa Rica survive? And how much climate pollution gets generated by long haulage of produce back to US consumers?

In fact, in the words of the international peasant network *Via Campesina*,

> Carbon trading has proven extremely lucrative in terms of generating investor dividends, but has completely failed in reducing greenhouse gas. In the newly invented 'carbon market' the price of carbon keeps dropping to rock bottom, which encourages further pollution. All carbon emissions should be reduced from the source, rather than allowing payment for the right to pollute.[7]

Conventional programmes for mitigating the collateral damage of consumer economies – melting icebergs, species loss, pollution-induced cancers – simply band-aid a competitive masculinist neoliberal system tailored to production for individual gain. In Australia, the Labor Party's lapsed Carbon Pollution Reduction Scheme (CPRS) was symptomatic of this *Androcene* reasoning, and it would have transferred a $13 billion compensation payment from people's pockets to polluter's pockets.[8] Injustice aside, emissions trading makes no sense from an ecological point of view. The approach arbitrarily prioritizes the carbon variable from a complex ecosystem, presumably because emissions are measurable, so can be priced. Against the solipsism of such economic reasoning, it is the functional integrity of natural metabolic processes that has real value, and emissions trading does nothing to preserve that. This is why peasant movements and reflective climate activists argue for the cancellation of carbon trading and offsets.

Embodied Debt

In the push for 'resource efficiency', ecological modernists externalize production costs on to the living bodies of others, then on to natural habitat down the line. Thus in the Eurocentric vision of a 'third industrial revolution', Germany as 'the responsible energy-efficient technician' is really living on credit, buoyed up by an increasing ecological debt for nature in the Global South, a social debt to exploited factory workers, an invisible embodied debt to women as reproductive labour and an intergenerational debt to young people.⁹ The commitment to a global economy as perfect mechanism relies on externalizing problems, displacing difficulties. Thus Worldwatch has named 'population' as a major cause of environmental degradation, and in fact, the climate debate often brings old-style environmental talk about the need for global population control back into fashion. This removes the need to examine capitalist overproduction and consumption in the industrial North, by placing responsibility for the climate crisis on to the bodies of politically voiceless women in the Global South: a genocidal thesis. What kind of arithmetic is involved in this correlation between emissions and population? If 60 per cent of humanity is responsible for only 1 per cent of carbon emissions, why talk about population?

Where is the grassroots movement call to remake 'the social contract'? Ecological modernist ideology assumes the assimilation of communities across all continents into the capitalist economic system. But since this conventional development model is what is responsible for climate destabilization in the first place, it is inconsistent to speak of climate mitigation and development in the same breath. To its credit, the International Trade Union Confederation (ITUC) recognizes the need for a new model, 'for developing countries not to repeat the mistakes of the past but to engage instead in a different development path, so as to help build the low carbon, climate resilient and socially-fair world we need'.¹⁰ What is missing, though, is an acknowledgement that so-called developing countries in the Global South have been on a sustainable low-carbon path for thousands of years. As John Bellamy Foster reminds us, it is capital that spreads what Marx called 'metabolic rift', damaging ecosystems and appropriating people's livelihood resources for the manufacture of profitable commodities.¹¹ Trade union thinking remains embedded in this Eurocentric history and tends to assume there is no other way of provisioning. Thus, despite calling for a new model, the ITUC climate statement clearly envisages the continuation of industrialization. In this future deal, capitalist and worker are united in their commitment to 'efficient' technologies, skilfully designed to remake nature less wastefully than in the past. The hope is that new processes and gadgets can prevent the biosphere from being shredded by mining and incinerated by manufacture.

This utopian thinking is routine in establishment sustainability circles where consultants argue the case for 'de-materialization'. But the dematerialization thesis is quintessential capitalism. In fact it is symptomatic of the classic fetishization of the machine named by ecofeminists as first premise of the patriarchal scientific revolution. However, the argument that sophisticated, digitally enhanced production will generate the same output using less material throughput readily succumbs to the Energy Returned On Energy Invested (EROEI) effect. That is to say, the full chain of costs behind manufacturing the new 'green' technology itself is rarely factored in, rendered invisible as these are by externalization on to othered sex/genders, races, classes, ages or species.[12] Under the United Nations Framework Convention on Climate Change (UNFCCC) schemes such as the Clean Development Mechanism (CDM) or Reduction of Emissions by Deforestation and Degradation (REDD) deal with pollution from industrial nations by funding offsets like 'carbon sinks' in tropical forests – othered people's livelihood areas.[13] The planned green jobs mentioned in the ITUC programme rest on an optimism that 'targetted investments and policies aimed at creating green and decent jobs in certain sectors, such as renewable energies, energy efficiency and public transportation can help us overcome the job crisis we are living through, and unions today are willing to convey this message to the world'.[14] For this reason, there is a pressing need for education programmes to equip unions, activists and others with conceptual tools for thinking through the patriarchal-colonial-capitalist mode of interaction with ecological processes. Among these reasoning skills is the Jevons Paradox, which indicates that reliance on economic production and market instruments in order to trickle down environmental benefits can only increase the material turnover of nature, fuel inputs and carbon outputs.[15]

In 2010, a *Zero Carbon Australia: Stationary Energy Plan* appeared with a claim that one square metre of solar mirror would generate the same amount of energy as 20 tons of coal. The project was endorsed by the International Energy Agency (IEA), by Sinclair Knight, Sandia, Lockheed Martin, Bechtel, Pacific Hydro and Leighton Holdings.[16] It was widely networked to align the political priorities of business, community and activist circles.[17] A mix of technologies was advocated from wind power to methane, but the centrepiece consisted in acreage of flat glass mirrors aligned to capture the sun for a concentrating tower generating steam at 550°C. Additions to the scheme included innovative urban building design, water recycling, automated waste collection and a possible spin-off in the manufacture of an electric car for export. At every plant site across the country, engineering for the *Stationary Energy Plan* involved a radical transformation of landscape by tree clearing, drainage and levelling. Accumulated mirrors across a large field are likely to function as massive radiant 'hot plate', with atmospheric impacts on rainfall and weather stability. Local river flows across the driest continent on Earth would be resourced for cooling. This Beyond Zero Emissions plan ultimately foundered as 'uneconomic', but as this account makes clear, there is no such thing as 'a free lunch'. Neo-Keynesian proposals are not the way to solve climate change; nor is capitalist collapse mitigated by marketing green technologies and 'generating' green jobs.

Strategy II

As the Manila-based IBON group have pointed out: the Eurocentric history of the humanity-nature nexus needs to be reconfigured.[18] 'Stopping extraction helps maintain . . . low carbon cultures by producing healthy ecosystems that provide communities with food, water, medicine and shelter.'[19] Only the labour of people working hands-on in the landscape can begin to repair the damage done by mining, deforestation, agro-industry, urbanization and manufacture. Indeed, it has been calculated that global warming projections could be reduced by 20 per cent if land clearing ceased. In countries North and South, commercial development projects like logging, dam construction or biofuel cropping destroy vegetation that serves as a 'biotic pump'.[20] What this means is that living plants function as cyclic 'heat valves' recoupling CO_2 emissions through water evapotranspiration and restoring local temperatures with rain precipitation. Vegetation also helps renew groundwater and fosters carbon sequestration in soil. Subsistence farming and nomadic gathering economies in the Global South protect this climate dynamic. Moreover, intact local water cycles stabilize the global water cycle and associated weather patterns. An understanding of processes which couple the water-carbon cycle offers an ecological approach to resolving the climate crisis – as well as decolonizing the unspoken social contract of global capitalism.

As the movement for an alternative globalization got together at Porto Alegre in 2001. Ecological feminists too argued the advantages of subsistence economies.[21] Yet while the ambitious World Social Forum idea set in motion a culturally reflexive North-South dialogue, it would have difficulty maintaining a politically unified agenda. Soon however, the looming climate crisis opened up a new opportunity for movement coalition. In response to the COP15 stalemate in Copenhagen, President Evo Morales of Bolivia stepped forward, calling for a Peoples' World Conference on Climate Change and the Rights of Mother Earth. This April 2010 meeting in Cochabamba, hosted by the country's indigenous peoples and women, was an attempt to interrogate the hegemony of capital, reframing climate politics with a twenty-first-century social contract. The guiding principle of the Bolivian approach was the Andean philosophy of *buen vivir* – Spanish translation of the Quechua term *sumak kawsay*; sometimes rendered as *bien vivir*; translated into English as 'living well'. *Buen vivir* shaped the preamble of the People's Alternative Climate Summit thus:

> We are all valuable, we all have a space, duties, and responsibilities. We all need everybody else. Based on complementing each other, the common wealth, organized mutual support, the community . . . develop[s] . . . without destroying man and nature . . .
>
> Within the Living Well framework, what matters the most is neither man nor money; what matters the most is life. But . . . the two development models, the capitalist and the socialist need rapid economic growth . . .
>
> . . . development . . . is now the leading cause of global crisis and the destroyer of planet Earth, because of the exaggerated industrialization of some countries' addicted consumerism and irresponsible exploitation of human and natural resources . . . The new models must begin by accepting there are fundamental limits to the capacity of the Earth to sustain us. Within those limits, societies must work to set new standards of universal economic sufficiency.[22]

The World People's Conference on Climate Change and the Rights of Mother Earth developed an impressive list of recommendations from workshops on the structural causes of climate change, historical responsibility and climate debt, mitigation, adaptation, financial provision, technology transfer, deforestation, agriculture and capacity building. Morales presented the conclusions at a special meeting of the UN and in a subsequent diplomatic round including a visit to the Vatican. The Bolivian UN ambassador Pablo Solon was central in this advocacy.

Cochabamba was an historic decolonial moment, drawing together an alternative North-South climate constituency. Even so, the Final Conclusions of Working Group 13: Intercultural Dialogue to Share Knowledge, Skills and Technologies pointed to an area of weakness in climate justice thinking.[23] Mitigation, adaptation, financing and technology transfer are key UNFCCC topics under Long-Term Cooperative Actions. But in addressing technology, the Cochabamba workshop largely contradicted its own overarching objective of *buen vivir* by legitimating the ecological modernization agenda. The capitulation set in at section F, 'Enhanced action on technology development and transfer', where an urgent need to 'catch up' with industrialized nations was stated.[24] Development and technology were assumed to be necessary to respond to climate change and both assumed to be neutral. The statement elided direct causal links between consumer economies and climate destabilization. The environmental, social and culturally homogenizing effects of the affluent North's exported technology transfer were also passed over.

Sadly, the viability of tried-and-tested local technologies and indigenous land management capacities bowed to the dated rhetoric of a twentieth-century ideal of a uniform global economy.

> Transfer of technology must fully compensate the loss of development opportunities due to the costs and technological demands to developing countries to live within a restricted atmospheric space. Poor countries face climate-related challenges to their development that were not faced by the developed countries in the process of their own development.[25]

The definition of 'poor countries' here is uncritically colonial, with development understood in an 'aspirational' opportunity. In the ideology of ecological modernization, the poor are characterized as unsophisticated victims and patronized falsely as unwitting contributors to the environmental crisis. To repeat the point, *per capita* carbon emissions from the predominantly rural South are far lower than those of the urbanized North. But Eurocentric notions of poverty and development are used in this document, as if they were unproblematic terms. The technology statement reads as if it had been written in the industrialized world, and indeed, UNFCCC texts may well have been adapted by Working Group 13 in the absence of a shared vocabulary for arguing local 'eco-sufficiency and *buen vivir*'.

Back to Dependency

This political subterfuge demonstrates an important lesson for decolonial and climate activists engaged in international negotiations. Section F of the official Cochabamba recommendations would go on to describe the stages of economic growth as follows:

> Sharing the complete technological cycle, namely enhancement, development, demonstration, deployment, diffusion and transfer of new and existing innovative technologies is urgent and essential to strengthening developing country Parties capacities in particular those listed in Art. 4.8 of the Convention. Developing countries must be recipients of the technological cycle in its integrity.[26]

The phrase 'technological cycle in its integrity' is mystifying. Transferred technologies are both out of sync with the environments in which they are manufactured and they result in a further loss of integrity in the environments they are exported to. Beyond this thermodynamic destabilization, the transfer of mechanized and digitized technologies takes a heavy toll on the symbolic integrity of daily practices in non-industrial cultures. This distribution of saleable products by transnational capital is assisted by bureaucratic agencies such as the World Trade Organization (WTO), the Global Environment Facility (GEF) and the Inter-Governmental Panel on Climate Change (IPCC), each one a tool of neoliberal mores.

The World Bank has also made a role for itself within the climate change establishment.[27] Thus, Cochabamba Working Group 13 conceded that 'Guidelines shall be established for the assessment and evaluation of technologies meant for transfer and deployment to ensure that they are environmentally sound and socially appropriate'.[28]

These instrumental guidelines hover above so-called developing communities as abstract forms of governance. They do not engage locally with people who oversee the humanity-nature metabolism on the ground. However, the Intercultural Dialogue to Share Knowledge, Skills and Technologies did offer this much:

> We recognize that indigenous and traditional knowledge and technologies form a valuable and useful part of the knowledge and technologies that are appropriate and useful for mitigation and adaptation activities in addressing climate change and that these have to be supported and be part of technology development, transfer and deployment.[29]

The question is: How are local knowledges and skills to be supported by capitalist financial instruments and bureaucratic regimes, whose very penetration of daily life unravels the coherence and practice of traditional knowledges? The contradiction is exacerbated once indigenous biodiversity expertise is classified under intellectual property law. It is noteworthy that CMPCC, 2010a, Clauses 47, 48 and 49 rejected private patents, demanding open access for all technologies. That seemed to speak to an intellectual commons, consistent with open access to livelihood resources like land, water and air, albeit puzzling given the general trend of recommendations.

However, instead of pursuing 'cross-cultural scientific dialogue' and the recognition of low-carbon subsistence economies in the Global South, the Cochabamba text read:

> We agree that early and rapid reduction of emissions requires the deployment of low-emission technologies on a massive scale and that developing countries particularly those with insufficient or no manufacturing capacity in environmentally sound technologies will have more difficulties in accessing adaptation and mitigation technologies and that measures shall be taken to facilitate and ensure their access to the technology.[30]

In Global North or South, the official line is that adaptation and mitigation can only be achieved through purchase or manufacture of new technologies. Thus, Working Group 13 adopted the dependency posture, rather than assuming global leadership by asserting the rationality of *buen vivir* with its small ecological footprint.[31] The demand for 'financing from developed country Parties amounting to at least 1 per cent of their GNP' may partly compensate the ecological debt incurred by Eurocentric plunder, but it effectively locks the South ever further into the capitalist machine.

By default, the conclusions of the technology group conceded to a transnational programme of neoliberal 'control' – one that echoed the now-defunct Multilateral Agreement on Investment (MAI) and reinforced the WTO. This arm of the modern patriarchal-colonial-capitalist imperium coordinates stakeholders at local, national and international levels. In this case it would create a Technology Executive Board; Technical Panels for adaptation and mitigation; Innovation Centres; and a Technology Action Plan. In addition, the Cochabamba meeting proposed a Multilateral Climate Technology Fund composed of Regional Groups of Experts in Investment and Development and a compliance mechanism to remove barriers to technology transfer, diffusion and development. While training was envisaged as top down, endogenous capacities were to be enhanced. An assessment of appropriate technologies was planned to look at economic and social factors as conceived by capital and its consultants, but here, cultural autonomy was rarely mentioned. Post-Cochabamba, UNFCCC documents reveal a bureaucracy that is consuming of energy, time and money, for governments, NGOs and climate activists that deal with it. The original Expert Group on Technology Transfer (EGTT) was wound up. The Subsidiary Body for Scientific and Technological Advice (SBSTA) and the Subsidiary Body on Implementation (SBI) would be complemented by a new Technology Executive Committee (TEC) and Climate Technology Centre and Network (CTCN), the latter to phase in at COP17.[32]

The pace of this evolving multilateral governance gives the lie to the urgency of climate change. Rather, it serves the morality of ecologically modern gentlemen, aiming to:

(b) Stimulate and encourage, through collaboration with the private sector, public institutions, academia and research institutions, the development and transfer of existing and emerging environmentally sound technologies . . .

(c) Develop and customize analytical tools, policies and best practices for country-driven planning to support the dissemination of environmentally sound technologies . . .

(iv) Stimulating the establishment of twinning centre arrangements to promote North–South, South–South, and triangular partnerships with a view to encouraging cooperative research and development.[33]

UNEP has become central in consultations with stakeholders in the Global South. The Global Environment Facility continues to conduct 'needs assessment' and fund technology transfer projects in conjunction with advice from business, the EGTT, UNEP, UNDP, UNIDO, the World Bank and UNFCCC, among others. The GEF will 'support technology centers and networks at global, regional, and national levels'.[34] Inside this non-transparent frame, the GEF would promote green capitalism with pilot projects like CO_2 Capture and Storage from Sugar Fermentation in Brazil; Green Trucks in China; and Renewable Wave Energy in Jamaica.

Metabolic Value

Against this surge of ecomodernizing coloniality, the main body of the Cochabamba Declaration reaffirms the need 'to recognise the plurality of forms of knowledge and ancestral practices, and transform scientific practices based on control over Nature toward paradigms oriented toward equilibrium with nature'.[35] But scarcely a trace of the original Cochabamba goal of *buen vivir* was carried forward in the UNFCCC negotiating text for the December 2010 COP16 in Cancun. Today, the transnational climate establishment and the grassroots global climate justice movement continue to walk parallel political paths. However, this sociological hiatus hands the peoples movement an opportunity to examine its contradictory thinking on technology transfer and finance for development. The government of Bolivia has remained ambivalent over economic development, even while endorsing food sovereignty and indigenous knowledges. It argues that technology transfer should be part of a climate debt owed by the Global North, free from conditionalities and Intellectual Property Right (IPR) restrictions. But ultimately, the Cochabamba Declaration had two faces – its second face an ecologically modernizing model, open for exploitation by business as usual.

Peoples of the Global South could be saying 'no thanks!' and modelling innovative political leadership at home – as do many ni-Vanuatu people from Oceania.[36] The eco-sufficient know-how of meta-industrial workers like peasant farmers and nomadic gatherers has much to teach old industrial communities and rust belts trying to establish sufficiency – would-be ecosocialists might take note of this. A development paradigm that functions in equilibrium with nature is also emerging from post-communist Europe. For example, at COP15, the People and Water NGO presented an integrative ecological strategy for climate stabilization.

> The living world influences the climate mainly by regulating the water cycle and the huge energy flows, which are closely linked with it. Natural ecosystems also develop in the long term towards the stabilization of closed cyclical processes (e.g. the water or carbon cycles), whose central medium is water and which efficiently manage solar energy with minimum material losses. Transpiring plants, especially forest growth demonstrate very efficient water management. They work as a kind of biotic pump, causing humid air to be sucked up out of the ocean and transferred to dry land.[37]

The uncoupling of water and carbon cycles that destabilises planetary weather patterns is a product of development: deforestation, industrial agriculture, urbanization and manufacture. Cleared land and paved cities draining water to the sea lead to a form of landscape entropy.[38] As underground aquifers dry out, the hydrological cycle is disturbed, soils cannot support plants or sequester carbon, which is then given up to the atmosphere as CO_2. Drying, devegetated land directly affects local weather because evaporative cooling of the air, cloud formation and rainfall are disturbed. Forests are much more than mere 'carbon sinks', and as climate expert Richard Betts of the UK Meteorological Office points out: 'the role of tropical forests in protecting us against climate change is severely under-rated.'[39] The Kosice Protocol, as argued by the People and

Water NGO, is scientifically verified, yet holistic rather than reductionist. It stems the metabolic rift of urbanization and agro-industry by encouraging human reciprocity with organic processes rather than control over them. In the villages of Slovakia, the People and Water NGO is motivating communities to protect their water catchments and their cultural identity in the land.

People working directly with nature, Indonesian peasants or Peruvian forest dwellers, for example, understand what a green job really means. A green job is one that regenerates ecosystems and human bodies through the creation of 'metabolic value'.[40] This reproductive form of economic provisioning points to the possibility of a climate-friendly alternative development model for the twenty-first century. In the words of *Via Campesina*:

> Sustainable local food production uses less energy, eliminates dependence on imported animal feedstuffs and retains carbon in the soil while increasing biodiversity. Native seeds are more adaptable to the changes in climate, which are already affecting us. Family farming does not only contribute positively to the carbon balance of the planet, it also gives employment to 2.8 billion people. Conversely, false solutions proposed in the climate talks, such as the REDD initiative (Reducing Emissions from Deforestation and Degradation), the carbon offsetting mechanisms and geo-engineering projects are as threatening as the droughts, tornadoes and new climate patterns themselves. Other proposals such as the biochar initiative, no till agriculture and climate resistant GMOs are the proposals of agribusinesses . . . It is unfair to use the benefits that small farmers provide to the environment as an excuse to keep polluting as usual.[41]

Eco-Sufficiency

The eco-sufficiency of 'living well' is a serious contender for the social ecological conversion of industrialized economies, but it means capacity building in a reverse direction – with peoples of the North listening respectfully to peoples of the South. The Civil Society Declaration on Technology and Precaution released in the lead up to COP15 reflected this positioning. Deferring to the

Cartagena principle, the Declaration outlined many shortcomings of ecological modernizing technologies:

> In many cases, action to address climate change is within our reach already and does not involve complex new technologies but rather conscious decisions and public policies to reduce our ecological footprint. For example, many indigenous peoples and peasants have sound endogenous technologies that already help them cope with the impacts of climate change, and to overlook these existing practices in favour of new, proprietary technologies from elsewhere is senseless.[42]

A long list of global climate activists signed on to this view, including Science for People; the African Biodiversity Network; Asian Women's Indigenous Network; Amigos de la Tierra, Costa Rica; Gender CC-Women for Climate Justice; Mangrove Action; Pesticide Action Network, Malaysia; National Farmers Union of Canada; Stop GE Trees, US; and the Third World Network, among others.

In Australia, the Climate Justice Network also expressed sensitivity to the sociological costs of international policy:

> we need to direct the economy and society to regenerative sufficiency, away from the productivist exploitation of natural resources (in particular fossil fuels). New norms of development are required to shift to forms of regenerative growth, and these norms must drive and underpin any 'direct action' program.[43]

In India, a National Forum of Forest Peoples and Forest workers explained:

> There is a climate crisis around and no amount of free trade, capital or technology will eliminate the roots of this crisis. You forget that the crisis has emanated from the way your society is structured – an edifice based on an unending desire for resources and a way of life that sees nature as an object of exploitation and extraction.[44]

In order to roll back the current ecological and financial crises, new historical actors such as these groupings, not to mention youthful voices from the most affected generation, must be heard at international negotiations.

> Any new body dealing with technology assessment and transfer must have equitable gender and regional representation, in addition to facilitating the full consultation and participation of peasants, indigenous peoples and potentially affected local communities.[45]

The androcentric focus on engineering infrastructure and the obsession with economic growth invert the thermodynamic order of nature, emptying out its living material flows – its metabolic value. In the language of ecological modernization, 'biogrowth' means the exact opposite of organic flourishing; instead, it refers to the amount of biomass taken up by the machine.[46] By contrast, the reproductive economy of meta-industrial labour catalyses vital matter-energy exchanges, a humanity-nature nexus in reciprocity. Against the ongoing dismemberment and commodification of ecosystems, an alternative model of development would be premised on the common sovereignty of energy, land, water and air. Templates for this already exist in many low-carbon economies of the Global South. The global Climate Justice Movement supports leaving fossil fuels in the earth; community control over production; reducing the North's overconsumption; localizing food; holding up indigenous rights; and reparation for ecological and climate debts to the South.[47]

Decolonizing initiatives like these provide reality testing for political actors in a Global North. And it is not only neoliberal ecological modernizers who objectify 'Nature' as a resource; instrumental rationality mars some ecosocialist traditions too. To recognize the logic of low-carbon societies is to show respect for the worldwide majority of workers, and this also makes good sense for creating movement alliances across continents. At this conjuncture, the Left has to give up trying to turn grassroots activists into clones of Marx's industrial proletariat. The era of factory socialism is exhausted; its traditional labour force is in disarray; and historical agency is unlikely to emerge from people who are disoriented by automation or offshore relocation of their jobs. Walden Bello is certain that ultimately, a big-picture climate strategy must call on government regulation to support of local economic sufficiency.[48] Bello calls this 'deglobalization', and it follows the democratic principle of subsidiarity. The only rational strategy for union workers today is to grow a wider labour identity, joining with women's, peasant, indigenous, youth and ecological movements. This is not to give up on the struggle with capital but to intensify it synergistically by joining together all movements for change—Workers, womens, indigenous and youth.

Ecomodernization policies impose ecological, social and embodied debts at the periphery of capital, but such externalities undermine people's resources and capacities for sustainable living. Where IMF-funded projects, WTO-mandated free trade or neocolonial UNFCCC governance structures disturb an established society-nature nexus, three things happen. First, people's livelihood resources are reassigned across to business; second, their locally appropriate knowledge skills are diminished; and third, cultural and personal identities are crushed.[49] However, in parts of Africa and Oceania wherever women are known to feed communities by means of low-impact subsistence farming, people are buffered from economic precarity – an insight that surfaced during the global financial meltdown. Development as understood by multilateral agencies is quantitative, whereas livelihood as understood by the commoner is qualitative – grounded in a functional relation with nature. A dollar a day thus has a different meaning for a Bangla Deshi farmer with access to land than it has for a bag-lady sheltering in the New York subway. Too many well-meaning professionals and activists miss this profound difference. It is time to make a clear choice – not so much between Right and Left but between strategies of ecological modernization and *buen vivir*.

8

Climate Science and Water

Coming to Our Senses

As George Bush senior told the world a few years back, the American way of life is 'not open for negotiation'. Yet consider the daily climate toll of transglobal supermarket supply chains using refrigerated transport and storage.[1] More significantly, the United States military is the planet's single greatest source of carbon emissions, as year by year it spends 'trillions of dollars waging war, chasing shadows, selling arms and fuelling conflict all over the world'.[2] Australia is another culprit. As we speak, a new federal Labor government is full speed ahead subsidizing gas exploration. Moreover, in partnership with Exxon, BHP, Rio Tinto, James Hardie, Boeing, Qantas, Mitsubishi, AGL and Origin Energy, it is funding a multimillion geoengineering project to save dying corals on the Great Barrier Reef from global warming. This 'best practice' Reef Restoration and Adaptation Plan (RRAP) involves scientists from Southern Cross University and the Institute for Marine Science in Sydney. However, in defiance of Biodiversity Convention clauses that outlaw environmental manipulation, the project will use barges to spray trillions of nano-sized ocean salt crystals into the air. The aim is to create thick white clouds reflecting sunlight back into space. Community locals are in disbelief. Cautious scientists calculate that the experiment will likely increase tropical rains and run-off while decreasing weather precipitation in Amazon forests across the Pacific.[3] What these patriarchal-colonial-capitalist decision-makers do not seem to understand is that the Earth is not simply 'an object' but a living metabolism.

Carbon Fetishism

Three decades ago, the UN Conference on Environment and Development (UNCED), also known as Rio Earth Summit, set up an inter-Governmental agreement on forest protection, albeit non-binding. Then in 2005, schemes for Reducing Emissions from Deforestation and Forest Degradation (REDD) were introduced. Yet as the World Rainforest Movement reports, moneys thrown at the climate problem through the World Bank's Forest Carbon Partnership Facility and private companies are ineffectual and socially disruptive. Most such remedial efforts are simply designed to buy carbon credits as a political trade-off for the production of emissions elsewhere.[4] The UN is often involved in green washing initiatives. Thus, by 2020, its policy was guided by a Collaborative Partnership on Forests with a Strategic Vision towards 2030 in conjunction with the Sustainable Development Goals (SDGs). Now protection was aimed at sound 'management' of forest 'goods and services'. Nevertheless, exploitation has not ceased and annual global forest loss is said to cover an area the size of Britain.

From the start, it was clear why activists renamed the Glasgow COP 26 in November 2021 – 'Conference of Polluters'. A genuinely ecological response to planetary destabilization would have focused less on Net Zero carbon and more on restoring self-regulating natural cycles that balance water-soil-biodiversity-carbon and weather. But the international meeting of governments chose to stay with counting emissions for use as offsets and even marketable derivatives. Brazil, Russia, India, China and Australia backed down on the goal of fossil fuel elimination by 2030. Simultaneously, the Glasgow Declaration on Forests, described by the UK, Colombia and The Nature Conservancy as 'unprecedented', simply rejigged old commitments around newly tradeable products. This occurs, as the planet's loss of evapotranspiration capacity diminishes year by year from forest clearing in tropical Brazil and Indonesia.

It is no wonder that international surveys of young people show them feeling abandoned by world leaders. Nevertheless, in May 2021, thanks to local youth, Friends of the Earth, Action Aid and others, The Hague ruled that Royal Dutch Shell had violated the 'right to life' and must reduce CO_2 emissions by 45 per cent within ten years.[5] This was an historic victory, if still only a small step along the path to climate justice. When BP admits that its use of shale gas will increase global emissions by 29 per cent by 2035, the international political battle over carbon emissions is far from over.[6] At the same time, the solution is held back by the reductionist, indeed fetishized, 1/0 methodologies of conventional engineering and economics. The debate is largely single-issue, fixated on the measurement and pricing of carbon molecules rather than the thermodynamic interplay of living ecological processes. Philosopher-activist Larry Lohmann describes this myopic focus on carbon as 'an endless algebra' and sees its peculiar construction of energy as 'blind racist colonialism', not to forget the sex/gender element.[7] Yet there are Other ways as hydrologist Michal Kravcik writes:

> some 58,000 square kilometres of paved urban paradise drains rainwater into the sea, leaving heat at a loose end in the atmosphere and causing chaos...
>
> [But] it would not be a paved road to hell, if we harvested rainwater in city parks and green areas, and let the water evaporate. And what's more, more water in the country means more vegetation, and more vegetation means more photosynthesis plus more consumed CO_2 from the air into vegetation.[8]

An ecological approach to climate interconnects water, soil, plants, sunlight, air and people. Landscapes are not merely passive sinks for CO_2 emissions but active agents of a life-affirming equilibrium in Earth temperatures. Already in 2010, a comprehensive literature review from *Current Opinion in Environmental Sustainability* was making this case:

> policy makers remain overwhelmingly focused on CO_2 reductions and continue to ignore other anthropogenic modifiers of climate systems. To

date, climate models such as those used for the 4AR of the IPCC have failed to adequately capture the full range of human-influenced climate forcings impacting on the climate system... It is critical to adopt a broader perspective ... [by examining] global and regional climate approaches which recognise the climate regulation function that forests and woodlands play through moderating regional climate variability, resisting abrupt change to existing climate regimes, as well as underpinning the hydrological cycle.[9]

The subsumption of abstract Nature by Human priorities – terraforming – is an enduring feature of *Androcene* ideology, and its linear preoccupation with 'counting' inevitably results in reductionist science. Moreover, there is a comfortable affinity between a factor being deemed measurable and therefore open to 'pricing'. Those who deal with the macro-measurement of single variables like CO_2 would be well served by a closer study of how ecosystemic energy flows are balanced by complex self-managing dissipative structures.[10]

Methodological Forcing

As things stand: patriarchal-colonial-capitalist entrepreneurs profit by selling 'solar renewables' to consumers looking for environmentally responsible products.[11] And the UN Framework Convention on Climate Change (UNFCCC) serves as a 'broker' for industry leaders hoping to export manufactured devices for climate adaptation into poor nations.[12] However, a thorough accounting of thermodynamic costs generated in making such products shows that tech fixes do not resolve but simply move costs around from one region of nature to another and from one class or generation to another.[13] When business as usual and climate catastrophists come together, the shared emphasis on hardware or end-of-pipe solutions often results in 'dangerous last-grasp strategies such as geo-engineering, nuclear and carbon markets'. Reinforcing this perception, scholars of science and technology studies, such as Kjetil Rommetveit, Silvio Funtowicz and Roger Strand, point out that:

> With increasing political, commercial and public pressures building up around climate science, the danger is increasing that hasty scientific conclusions feed into policy processes demanding fast and safe answers.

Policy makers and scientists may jump to premature conclusions leading to locked-in situations where society is committed to solutions that are neither sustainable, nor scientifically, nor economically viable.[14]

The construction of scientific knowledge never takes place in a social vacuum, as Thomas Kuhn explained decades ago.[15] Scientific facts and models are determined consensually, as informed researchers argue and decide among themselves what is plausible and what is not. But science as a democratic enterprise is readily distorted by corporate and government interference. This is evidenced in the manipulation of academic research funds, in pressures on government regulatory committees and in the censorship of investigatory journalists. The 'debate' over climate change is taking place in this 'social pressure cooker' context, and it is important that people be aware of that, as they rely on the integrity of scientists and on the capacity of state agencies to protect them from the risks of modernization. If ecological complexity can play havoc with predictive models, social interests can undermine them too. Thus, the practice of environmental risk management is fraught with difficulty and the more so, because it involves balancing multiple physical and social variables and time scales at once. Where this scientific uncertainty is considerable, scientists will sometimes protect themselves in advance against incrimination for failed risk analysis, by engaging community members as co-authors. Similarly, governments, these days, secure accountability by inviting the broadest possible public constituency to participate in policy deliberation. However, uncertainty and accountability problems are magnified in the case of the Inter-Governmental Panel on Climate Change (IPCC), because its mandate is to judge variables on a global scale – and politically speaking, the globe is not 'a level playing field'.

Some scientists engaged in the UN processes have expressed uneasiness about such pressures. A survey carried out by Ann Henderson-Sellers, former director of the World Climate Research Programme in Geneva, lists the following epistemological concerns shared by lead authors of the Fourth Assessment Report of the IPCC in 2007:

1. the need for complexity in modelling

2. the need for a fuller understanding of the carbon cycle

3. the need to recognize links between land-use change and greenhouse emissions
4. the need to rectify geographic unevenness in existing climate data
5. the need to include measures of the hydrological cycle
6. the need to bring social and economic sciences to the analysis of climate[16]

Several research scientists from Eastern Europe have been path-breakers in integrative thinking about the microphysics of solar energy dissipation and temperature control. Led by Jan Pokorný, the team from University of Southern Bohemia, Mendel University, the Czech Life Sciences University in Prague and the ENKI research organization states:

> Ecosystems use solar energy for self-organisation and cool themselves by exporting entropy to the atmosphere as heat. These energy transformations are achieved through evapotranspiration, with plants as 'heat valves' . . . While global warming is commonly attributed to atmospheric CO_2 . . . it is critical that landscape management protects the hydrological cycle with its capacity for dissipation of incoming solar energy.[17]

Among this group, the late Wilhelm Ripl, an Austrian limnologist, looked at how landscape interference loses fertile carbon matter to lakes and seas, with entropy the result. Ripl concluded that the so-called 'developed' landscape simply replaces order with randomness.[18] Theorists Anastassia Makarieva and Victor Gorshkov from the St Petersburg Nuclear Physics Institute would challenge common assumptions about atmospheric water circulation, again affirming the critical ecological agency of plant life.

> The intense condensation associated with high evaporation from natural forest maintains regions of low atmospheric pressure on land. This causes moist air to flow from the ocean on to land, compensating for continental water loss through river runoff. Conversely, deforestation induces desiccation by reversing this moisture flow . . . forest preservation is a sound strategy for both water security and for protecting a continental landmass against climate extremes like floods, droughts, hurricanes, and tornadoes.[19]

Scale Versus Responsibility

A major obstacle to integrative thinking on climate has been a lack of articulation between instruments of international governance like the Kyoto Protocol, the Convention on Biological Diversity or the Millennium Development Goals. Indeed, Harvard scholar Sheila Jasanoff calls for new institutions altogether, allowing for interaction between scientists and citizens:

> the very fact that judgement has been integrated across so many fields leaves climate science vulnerable to charges of group think and inappropriate concealment of uncertainties ... Though intergovernmental in name, the IPCC is subject to none of the legal or political requirements that constrain, but also legitimate, national expert committees ... IPCC performs a mix of functions – part scientific assessment, part policy advice, and part diplomacy – that demand external, as well as internal accountability.[20]

Jasanoff's point is well made, for most concerned citizens assume that the IPCC is conducting 'pure science' – not acting out a blend of political roles: 'part scientific assessment, part policy advice, and part diplomacy'. It is hard to imagine how research into the complexities of climate can flourish under such circumstances. In sociological terms, global warming is the collateral damage from a business-driven industrial growth trajectory with ever-increasing demand for natural resources, cheap labour and consumers. With the globalization of economic production and intensification of free trade, nation states begin to cede powers to supranational institutions like the World Trade Organization (WTO) and UNFCCC.

A political consequence of this gradual shift to international governance is that people lose control over their everyday conditions of existence. The democratic ideal is disconnected from community. At the same time, the capitalist division of labour, trained specializations and abstract expertise, along with the growth of urban consumer lifestyles, all disconnect people from a direct sensuous understanding of how material nature works and how their very own bodies are continuous with that same nature. Commenting on the consequences of Eurocentric modernity, philosopher Stephen Toulmin writes:

> There is a . . . contrast between our local knowledge of the patterns we find in concrete events, and the universal, abstract understanding . . . The substance of everyday experience refers always to a 'where and when': a 'here and now' or a 'there and then'. General theoretical abstractions, by contrast, claim to apply *always* and *everywhere* – and so . . . hold good *nowhere-in-particular.*[21]

If findings apply 'no-where-in-particular', then do the risks of scientific uncertainty also apply 'no-where-in-particular'? In other words, is there an inverse relation between scale and the capacity for responsible science? And how do scientific results applicable to no-where-in-particular translate into responsible social policy? How do people struggling to protect their health, livelihood, community and habitat make good use of scientific results that apply 'no-where-in-particular'?

Then again, what does it mean to talk of 'acceptable risks' in an international context? Are acceptable risks simply those that can be displaced on to othered humans in othered environments? As the Peruvian ecofeminist Ana Isla has demonstrated, the livelihood of subsistence dwellers in forests of the Global South is sacrificed when pollution from wealthy industrial nation states is 'offset' by carbon sinks.[22] This occurs under the UNFCCC Clean Development Mechanism (CDM), and according to climate justice activists, ecological debts of this kind are inherent to programmes for Reducing Emissions from Deforestation and Degradation (REDD). A sociologically reflexive, non-sex/gendered, non-racialized UNFCCC policy would ensure that wealthy states deal with their own CO_2 footprint on their own territory. As distinct from the politically convenient gaze of no-where-in-particular, a democratic approach to climate change will look at things from the ground-up. It will apply the principle of subsidiarity and respect the localized experience and knowledge of those who labour everyday to maintain living processes – as small farmers, mothers, gatherers and fishers do.[23] Strong sustainability means empowering environmentally committed local communities, and this is the focus of the alternative globalization movements of movements.[24] Exemplary here is the South African Climate Justice Charter, initiated by the Cooperative and Policy Alternative Centre for Systemic Alternatives (COPAC). The charter integrates transitioning to food and water sovereignty in peoples' assemblies of Earth citizens – Interfaith, Labour, Indigenous, Women and Youth – committed to a deep ecological sense of intrinsic value in the web of life, the Rights of Nature, communal ownership and local governance.

It is time to round out the study of climate with more holistic ecological research and sociological analysis. For example, science is increasingly embedded in the culture of business as usual. This subtle shift from twentieth-century notions of scientific neutrality and 'objectivity' is revealed by Rommetveit's concern over climate being turned into an object of international agency management. How objective is the treatment of local ecosystems in all their idiographic uniqueness, once the political decision is made to go for global as distinct from regional assessments? Are 'methodological forcings' now introduced? Can data gathered in differently functioning ecosystems be treated additively? Does the prioritization of computer modelling over careful, on-the-ground empirical observation create further methodological forcings? Computer simulation may work for industrial processes where humans are in charge of inputs, but the dynamic couplings and oscillations of nature's metabolism may well evade easy prediction.

Peoples' Science

Current international climate assessments are based on abstract, decontextualized, global averages, but as social scientists suggest, that kind of methodology appears to be tailored to serve a social engineering agenda. Any top-down process is undemocratic. But additionally, in the context of climate change, the richness of scientific findings is compromised, if people with a diversity of skill sets and observations are excluded from the activity of knowledge building. For example, once it is appreciated how the carbon and hydrological cycles are interlocking, and regulated by plants as 'heat valves', it makes sense to bring the hands-on land management expertise of farmers or indigenous forest dwellers into deliberations over climate mitigation. Ideally, the composition of committees like the IPCC would have a balance of class, race, sex/gender and youth membership.[25] A too narrow social base will foreclose a panel's terms of reference, choice of methodologies, attitudes to uncertainty and risk. The Manila-based Water for the People Network offers an exemplar of grassroots water management skills.[26]

In defiance of the *Androcene* compulsion that turns organism into machine, Kravcik uses the phrase 'hydraulic mission' to explain how nature's feedback cycles have been broken by the ancient drive to mastery and control of water. His work on restoring rural catchments in Slovakia has evolved into grassroots education and employment projects, benefitting community development, even enriching cultural identity. So he writes,

> it is deforestation, industrial agriculture, and urbanisation that determine climate by draining land, so that more solar energy re-enters the atmosphere as sensible heat, rather than latent heat of evaporation. Human made 'hot plates' lead to irregular precipitation and other climate destabilisation effects, but these can be mitigated through rainwater conservation and re-vegetation. This integrative paradigm combines the management of climate, water, biodiversity, and land, with implications for agriculture, forestry, engineering, urban design and regional planning.[27]

In Australia, Peter Andrews and Duane Norris devised a parallel technique called Natural Sequence Farming to restore landscape hydrology. As they say, agro-industrial farming 'mines carbon' from the soil, whereas NSF enhances metabolic growth and soil fertility, achieves aquifer recharge by avoiding erosive run-off, avoids compaction by hooved animals and gets the water and carbon cycles back into sync.

> Early settlement of the continent by people with European cultural assumptions disrupted established interactions of water, soil, and plants resulting in lost fertility. Moreover, [the continuation of] agricultural practices such as clearing, burning, ploughing, draining, and irrigation, have implications for global warming. Soils hold twice as much carbon as the atmosphere, and three times as much as vegetation. But carbon in exposed soil oxidises releasing CO_2 into the atmosphere.[28]

Getting the science right is certainly necessary but it is not sufficient. Failures in administrative coordination show why the social antecedents of climate policy cannot be ignored. Taking action on climate means acknowledging the full reach of the patriarchal-colonial-capitalist imperium:

- the need for 'management and control' inherent to patriarchal culture
- the respective roles of coloniality and business as usual
- the reliance on reductionist 1/0 reasoning in science, governments and agencies
- the blind faith in man-made technological solutions for ecological problems
- the externalization of risk by cost shifting on to Others without a political voice
- the lack of sex/gender, race, class and generational reflexivity in decision-making.

An encouraging outcome of the climate crisis is that it has given ordinary people across nations and classes a sense of themselves as global ecological citizens. One challenge, however, is getting past the feel-good macho clash between sceptics and believers.

Moving On

As Exxon was said to be spending lavishly on public relations firms like Edelman, the Inter-Governmental Panel on Climate Change (IPCC) renewed its call to action in the *Sixth Synthesis Report*.[29] The warnings were familiar: without an urgent reduction in fossil-fuel use by industrial nations, average global temperatures are estimated to rise by 1.5 degrees above pre-industrial levels in the first half of the next decade. The alternative is heatwaves, floods, melting ice, seawater rise, drought, crop failure, malnutrition and species extinction. The emission of heat-trapping gases like carbon dioxide and methane actually increased in 2022, so fossil fuel usage needs to be halved by

2030 and cease by 2050. At the 2022 meeting of COP 27 in Sharm el Sheik, this point was edited out of the final agreement under pressure from the oil producers. Shell even argued that the Net Zero goal cannot be realized this century. The UN urges further expenditure in clean energy generation by solar and wind; yet reliance on battery storage, promotion of electric vehicles and further digitalization of lifestyles means ever more land degradation from digging for heavy minerals like lithium iron and nickel graphite. Meanwhile, climate activists were expressing concern over facilities for public input at the planned 2023 meeting of COP 28 in Dubai. And indeed, this meeting turned out to be another 'go slow' on fossil phase out, with 2,456 company lobbyists present and COP president Sultan al-Jaber using the United Arab Emirates summit to secure new oil deals.[30] Al Gore was equally disappointing, given his advocacy of 'the nuclear alternative', and in general, a lot of weasel words about methane, as well as carbon capture and storage, were exchanged. The Australian Labor government is said to have come to the table with over 100 new mining applications in its back pocket. The horror of Small Pacific Island States was palpable over evasion of the liability question and an inadequate Loss and Damage Fund.

In contrast to the futility of COP 28, the 2023 United Nations Water Conference indicated an interest in exploring the holistic water-for-climate approach.[31] To paraphrase the tentative optimism of former Bolivian diplomat Pablo Solon: just as the UN was designed to replace the League of Nations after its failures in the Second World War, so a radical reform of the climate negotiation process might be on the cards . . .[32] *La lutta continua.*

9

A Just Transition?
Women Are the Key

In 2020, the Institute for Global Development (IGD) at the University of New South Wales, Sydney, held a workshop on Gender and Just Transitions for women activists from diverse backgrounds. Here is an edited exchange on ecofeminism between Somali Cerise, IGD Research to Practice Associate and myself.

> ***Somali***: *At our workshop on Gender and Just Transitions you emphasized that a just transition requires fundamentally rethinking the relationship between human beings and nature. This means we stop seeing the environment as there simply to serve human interests and, instead, view humans as just one part of the ecosystem. Can you elaborate on why we must move from the West's anthropocentric dualism of 'Humanity over Nature' to achieve a gender just transition?*
>
> **Ariel**: Yes, this broader public understanding of the global economy is critical to survival of Life-on-Earth. Crises like climate change, biodiversity loss and the 2020 pandemic are each outcomes of the dominant imaginary that positions Humanity versus Nature. This dualism, derived from ancient Abrahamic religious cultures, would be secularized by the European Enlightenment and scientific revolution. Modern science now shifted from seeing nature as a living organism to

Nature as a 'machine' that could be designed and improved by men. The logic of Humanity over Nature also implied Subject over Object, Mental over Manual, Production over Reproduction, Man over Woman, White over Black. This life-alienated objectifying ideology is closely tied into Eurocentric masculinist practices that are indispensable to colonial and capitalist systems. It is not only women who are conventionally treated as 'closer to Nature than men' but also indigenous peoples and children. This preconscious hierarchy of capability, entitlement and power infuses everyday talk and political decision-making.

Andro Alienations

Most governments and multilateral agencies are now taking the environmental crisis seriously – the international Anthropocene conversation is a marker of that. Yet the very term Anthropocene is part of the problem, since the mainstream political discourse itself is anthropocentric. Academic disciplines, economics and law are all premised on the superordination of Humanity over Nature. But this cultural lens blurs the fact that the choices, decisions and actions of subjected populations derive from masculinist values. In fact, we should be calling this era the *Androcene*.

As an empirical fact, all humans are nature – simply 'nature-in-embodied-form'! People involved in the labour of nurturing young bodies or growing their own food know this very well. So it was that five decades ago, women opposing polluted urban neighbourhoods in the Global North or local deforestation in the Global South came to recognize the destructive arrogance of the dualist 1/0 logic and its instrumental reasoning. Working with natural processes means facilitating living metabolic transfers, so learning complex skills and the necessity of a precautionary ethic. From this vantage point, the social and ecological crises that we experience today are an effect of attitudes, embedded in the sex/gendered political economy of international institutions.[1]

The politics and theoretical literature of an ecological feminism developed from this insight. Ecofeminists also noted how in patriarchal-colonial-capitalist societies, the resourcing and commodification of nature occurred in parallel to the resourcing and commodification of their own generative reproductive bodies. The latter exploitation can be seen today in the existence of two parallel sex/gender paradigms Public over Private: an individualistic monetized economy (ME) and a non-monetized relational economy (WE).[2] The domestic WE economy materially maintains the ME economy but is generally treated as a 'natural' activity.

> **Somali**: *What does ecofeminism propose as an alternative to the dualism of 'Humanity versus Nature'? What would be some positive examples that we can learn from?*
>
> **Ariel**: Ecofeminist activism for Life-on-Earth responds to the interconnected injustices of neoliberalism, militarism, corporate capture of science, worker alienation, reproductive technologies, sex tourism, child molestation, neocolonialism, extractivism, nuclear weapons, land and water grabs, deforestation, animal cruelty, genetic engineering, climate change and the Eurocentric mythology of progress.

Embodied Thinking

Ecofeminist thinking offers an alternative epistemology, a way of knowing quite distinct from the manipulation of people and nature. Yet it would be ideological nonsense to attribute women's political insights to some inborn 'feminine essence'. The source of ecofeminist judgement is neither biological embodiment nor cultural mores, although these will influence what is perceived. Rather, the source of an ecofeminist way of knowing is labour, as people discover understandings and skills through interaction with the material world. People like caregivers, farmers and gatherers are in touch with all their sensory capacities, so able to construct accurate and resonant models of how one-thing-joins-to-another. This is an 'embodied materialism'.

The global majority of women as caregivers or reproductive labour have been historically positioned right at the ontological margin where so-called Humanity and Nature meet. Unlike factory or clerical workers, culturally diverse groupings of women oversee biological flows and sustain matter/energy exchanges in nature. In fact, the entire thermodynamic base of capitalism rests on material transactions mediated by the labour of this unspoken meta-industrial class. Day by day, the global economic system is accruing a vast unacknowledged debt to these workers. In recent decades, women caregivers in the Global North and colonized communities in the South have come together in a political movement of movements charged by the knowledge that emancipation and sustainability are interlocking goals. The unique rationality of their meta-industrial labour is a capacity for economic provisioning without externalities – that is to say, without passing on a social debt to others or forcing natural processes into degradation and entropy.

- In Ecuador, women of *Accion Ecologica* (www.accionecologica.org) have invented a concept of 'ecological debt' to describe the 500-year-old colonial theft of natural resources from their land – the ongoing modern theft of World Bank interest on development loans.

- In the United States, Code Pink activists work tirelessly for world peace; others focus on ending cruelty to animals.

- In Africa, women whose livelihoods are threatened by mining near their village homes have established WoMin (www.womin.org.za), a continental anti-extractivist network with its own ecofeminist manifesto on climate change presented to COP25 in Paris.

- In China, many village women are refusing to use industrial fertilizers and pesticides, choosing to restore soil fertility by reviving centuries-old organic farm technologies, then modelling communal food sovereignty.

- In India, the *Navdanya* (www.navdanya.org) network organizes schools for eco-sufficiency and 'banks' traditional seeds to save them from biopiracy and corporate patenting by Big Pharma.

- In Australia, suburban housewives known as MADGE actively oppose genetically engineered foods. Others known as the Knitting Nannas are fierce protectors of the east coast river network.

- In France, young women and men are pioneering economic degrowth and rebuilding vibrant communities around permaculture.

Somali: *What are the different roles of actors – from governments to social movements – in achieving this shift?*

Ariel: At Rio+20 the business sector, politicians, World Bank and UNEP stepped up with a Green New Deal proposition. This was later exposed as a public relations exercise for an emerging nanotech-based bioeconomy. The patriarchal-colonial-capitalist method of protecting what they call Nature is to commodify 'ecosystem services', subsuming the living metabolic flows of forests, sunlight or ground bacteria, under a pricing mechanism. Similarly, the International Monetary Fund and others advance a Green Economy built on free market ideology. But intellectually, decision-making by world leaders relies on a thoroughly incoherent, and thus publicly confusing, vocabulary of 'financial capital', 'human capital', 'natural capital' and 'physical capital'.

Global Actors

Many well-meaning citizens in both Global North and South believe technology transfer and digitalization is necessary to achieve 'a just transition' to sustainability. The preferred androcentric response to the crises of globalization is innovation. It is claimed that new technological efficiencies can de-materialize the amount of resources used by industry. However, automated production does not avoid displacing self-sufficient rural communities for mineral extraction, nor does it avoid heavy energy drawdowns for manufacture. As I keep on saying, the engineering 'optimization' of material throughput rarely factors in all the relevant operational aspects of mining, smelting, manufacture, communications, transport and waste disposal. When fully researched, ecomodernist expectations of progress do not hold up. The only thing that holds up through the rhetoric of scientism is the illusion of mastery and control.

UN agencies and some NGOs which adopt the Eurocentric liberal political discourse tacitly sanitize environmental, decolonial or sex/gendered matters by departmentalizing them as separate 'single issues'. This piecemeal problem-solving policy inadvertently disguises existing, often intersectional, power relations because it stops people 'joining the dots'. For sure, micro loans are offered to poor Bangladeshi women, but this is hardly liberating. As long as thinkers at the UN are guided by liberalism, and people are processed through an objectifying divide and rule formula as 'stakeholders', progress towards 'a gender just transition' will be very slow. Contrasting the Women's Beijing Plan of Action and the UN Millennium Development Goals (MDGs), Caribbean feminist Peggy Antrobus diagnosed the MDGs as 'Most Distracting Gimmicks'!

> **Somali**: *What would a different system of power and economic relationships look like?*
>
> **Ariel**: Well to underline again two astonishing facts: the global economy already overshoots planetary capacities by 50 per cent every year; and the Sustainable Development Goals (SDGs) do not remedy this.[3] The World Bank and the UN SDGs promote privatized management of water supply. But since markets can only increase the value of a commodity by making it scarce, this method of water protection is a contradiction in terms. Similarly, environmental solutions like carbon trading, geoengineering or climate smart agriculture will not restore nature's life-support systems once these are continually being broken apart by technological adventurism. For small producers, landless rural women and indigenous peoples, the Green Economy is just another structural adjustment programme deepening their precarity by realigning national markets.

Regenerative Models

In response to imposts such as these, a global movement of movements firmed up after the 1999 Battle for Seattle against the World Trade Organization. This broad people's alliance held its first World Social Forum in 2001 – and two decades later, WSF members, albeit in process of reorganization, still believe Another Future Is Possible! On the streets of Davos outside the World Economic Forum and at the UN COP negotiations, activists pursue political subsidiarity and eco-sufficient commoning.

The worldwide Rosa Luxemburg Stiftung is also working on Socio-Ecological Transformation. Yet in both Global North and South, among elites and cadres alike, there is a need for 'capacity building' to include more effective sex/gender consciousness-raising across the political movements. It is time to hear, reflect on and activate women's critique of the *Androcene* imaginary and how it consolidates its powers through patriarchal-colonial-capitalist institutions. Earlier forms of feminism – liberal, socialist, postmodern – still had androcentric framing; whereas ecofeminism, born in environmental struggle, was oriented to *oikos* from the start. As such, the analysis of ecofeminism was immediately transnational, cross-cultural and decolonial in focus. The ecofeminist subsistence model encourages European moves towards degrowth, South America's *buen vivir*, India's *swaraj* communities, the South African ethic of *ubuntu*, Oceania's *kastom ekonomi* and the food sovereignty goal of Via Campesina.[4]

As noted, a marked sociological divide continues to exist between the middle-class, white, masculinist culture of business, governments, multilateral agencies and transnational technocrats versus those whose livelihoods are destroyed by industrializing development models, inconsistent climate policy, militarized resource grabs and, even closer to home, by domestic violence. Ideally, a just transition will be guided by a precautionary ethic and maintained by regenerative skills.

10

Food Sovereignty

Meeting Real Needs

As the social and ecological costs of global free trade add up, and nation states serve corporate profit rather than people's needs, alternative models of provisioning emerge from unexpected quarters.[1] In parts of China, for example, women are showing how 'food sovereignty' might be achieved hand in hand with environmental flourishing. Their 'real green jobs' prefigure business, UN and government calls for a Green Economy. Earlier Maoist ideals of communalism in farming, heath, education and welfare began to disappear in the 1980s as Premier Deng Xiaoping made neoliberal reforms and sought to position the country among major international powers. The move to economic competition lives on in public motivational slogans like 'Democracy is a means of Entrepreneurship!' and 'Let's All Get Rich Together!' The outcome was that China's surplus in GDP terms would be phenomenal, signified in the growth of elegant cities like Hangzhou and Chongqing. But many regional cities funded public housing and welfare programmes by 'rationalizing' rural settlement, and invariably land speculation followed. A new class dynamic began taking shape, with excessive investment in urban high-rise construction – now shown to have been a hollow choice as China's real estate market falters. The inherent risks of capitalism as a socio-economic system do not really fit well with 'development and progress'. By contrast, traditional Chinese agriculture techniques have regenerated land and peoples for over 4,000 years.

Internal Colonies

What began in late-twentieth-century China was a form of internal colonization, or primitive accumulation, much like the history of enclosures in Europe. But whereas that post-feudal class structure took shape over hundreds of years, in China the process was compressed into a few decades. With 'market socialism', government at all levels designed opportunities to encourage a new middle class of entrepreneurs. Party officials, often with business partners in tow, developed inducement schemes for peasants to give up smallholdings. In turn, the recruitment of once self-sufficient landed farmers fostered an industrial working class dependent on consumerism for survival. There are international media reports of migrant workers claiming to have been cheated in land deals. In China's cities, these now landless factory workers encounter indifferent labour conditions and few citizenship rights. Families are broken apart as children stay back in rural villages with grandparents. Central government subsidies to privately owned factories have resulted in local air, water and soil pollution. And as people's livelihood resources are turned over to industrial parks and export-oriented manufacture, workers face rising food prices. At least, if global financial crises destabilize urban employment in China, workers can return to the buffer of village life – although this once self-sufficient agricultural nation is heading for a future where only a quarter of its population will live in rural areas. The official solution is to lease overseas farmland to meet domestic food needs, a model well underway in parts of Africa. Pollution from supply chains, shipped or airfreighted between continents, is simply 'cost-shifted' to the planet globally. Meanwhile, inside industrializing China, fallout from toxic factory emissions affects the quality of farmland and community health. Scholars at the Chinese Academy of Science (CAS) report that of 11,590 grain crop varieties planted in China in 1956, only 3,271 varieties remained by 2014.[2] Another consequence of China's rapid industrialization has been population decline following the one child per family policy a couple of decades back. The government is now seeking to remedy this situation but not without cost to the present generation of young women who are both encouraged into 'motherhood' and facing restrictive university entrance requirements.

These material imposts of modernization on biodiversity and social relations are transferred from the centre of institutional power to the periphery. The extractivism is thermodynamic as much as it is economic: and it follows the androcentric hierarchy observed elsewhere: a social debt to inadequately paid workers; a livelihood debt to peasant smallholders; an embodied debt to women family caregivers; a generational debt to youth; an ecological debt carried within nature at large. This 'uneven development' threatens to erupt in conflict. Yet while urban workers organize around exploitation, a new interest in local livelihoods is taking off among village cooperatives and credit unions, NGOs and intellectuals. The movement for rural reconstruction based on traditional knowledges and skills is widespread, but an especially notable initiative is the practical learning centre known as Little Donkey Farm on the outskirts of Beijing. Another source of inspiration is PeaceWomen activists Lau Kin Chi and Chan Shun Hing, from the Global University for Sustainability in Hong Kong.[3] The Maoist revolution introduced agricultural modernization with the promise of high yields through hybrid seeds, artificial fertilizers and pesticides, animal antibiotics, irrigation and electrification. Three decades later, privatization policies would revive family-based production, albeit continuing the agro-industrial 'Green Revolution' approach of government technical advisors. But many Chinese women villagers are refusing the ecological and medical violence of chemical farming. As they say, 'nurturing the land is like nurturing a child.'

In mainland provinces like Yunnan, Sichuan, Shanxi, Hebei and on the islands of Taiwan and Hong Kong, women are setting up organic farming cooperatives. Bypassing the cynicism of government experts and farmer husbands too – low-intensity patriarchalism – these women of China seek the exhilaration of embodied labour. Their resilient polycultures work in reciprocity with nature. They develop animal-based bioliquid sprays for orchards and paddy fields; they give up weeding; they leave ground covers in place to encourage water retention; and they compost green manure to re-energize soil organisms. They preserve groundwater quality by avoiding microplastics from packaging, by converting waste oils to soaps, and they protect catchments by opposing dams. Alongside subsistence produce such as rice, oil, fish, chicken, pork, vegetables and fruit, plant medicines are promoted, as well as recyclable handicrafts like woven slippers, fishnets or bamboo furniture. Some women manufacture and sell pure soy sauce and vinegar; others make weekly household food deliveries to their communities by bicycle or van. As these pioneering women, many of their projects city-based, meet to compare crop yields and soil fertility, they build analytic skills and self-empowerment through cooperative learning networks. Public outreach follows, with workshops and street stands to educate passers-by on the need to replace toxic industrial agriculture with clean local eco-sufficient provisioning. Then there are the cultural benefits of this place-based commoning. One group runs a 'happy kitchen' serving 'slow food' to students while honouring the names of workers who grew it. The aim is to restore a sense of succession, celebrating the vernacular science of earlier generations and passing this on.

This philosophy is at once ecological, feminist, decolonial, socialist and intergenerational joining the logic of sustainability to the logic of community building, economic equity, cultural autonomy and peace. Like ecological feminists on many continents, these women offer a grounded understanding of peace and security and an opportunity to heal masculinist forms of socialism. Their politics runs deep. Its first premise is that political wisdom and strength 'grow from everyday life' and from being part of 'the cycle that never stops'. Livelihood networking has been encouraged through South-South Forums organized by the Global University for Sustainability and as organizer Sit Tsui from the Centre for Rural Reconstruction in Chongqing observes, the roots of 'food sovereignty' in China have a proud history among visionaries going back at least a century. At the same time, she points to the decolonizing power of the model as an alternative to modernity.[4] In this sense, the movement for rural reconstruction and food sovereignty in China dovetails with radical international initiatives to the failing global economy. These include the Landless Peoples' Movement in Brazil, Nyeleni in Mali and Via Campesina worldwide. They dovetail with practices like commoning, solidarity economies, bioregionalism, permaculture, degrowth, urban community gardens, the subsistence perspective and *buen vivir*. Here is a climate-savvy 'green economy' and 'green jobs' in the true sense of the word. It resonates with already-existing grassroots indigenous and peasant knowledges across Asia, Oceania, Central and South America, parts of Africa and the Middle East.

Benefit Sharing

The South China Agricultural University and the Chinese Academy of Science (CAS) also support organic farming initiatives. In regions like Qinghai, Yunnan, Hebei and Inner Mongolia, the circular agricultural model of 'maize-pig-and-vegetable' is particularly popular among farmers, because it meets subsistence needs simultaneously with natural fertilization, infestation control, biogas digestion and ecological benefits. Again, women lead in what has become the Farmers Seed Network, among them Lu Rong Yan, Li Ruizhen and Zhang Xiuyun.

> The FSN brings together over 30 communities as action pilots and living labs from 10 provinces across China and works closely with CCAP and the United Nations Environment Programme – International Ecosystem Management Partnership (UNEP-IEMP) of CAS, GMRI, Institute of Crop Science of the Chinese Academy of Agricultural Sciences (ICS-CAAS), China Agricultural University, Yunnan Agricultural University, Kumming Institute of Botany (KIB) and some civil society organizations, such as the Beijing Farmer Market and the China Community-Supported Agriculture coalition.[5]

With backing from the National Peoples' Congress, the Chinese Academy of Science has been promoting action research on Participatory Plant Breeding in rural villages since the turn of the millennium. It has been encouraging farm women's cooperatives in local resource collection and in community experimentation and registration of germ plasm. Today, a national conservation system consists in some thirty or more seed bank sites across the country. CAS describes farmer-led *in situ* community seed banks and scientist-led gene banks as complementary and 'mutually reinforcing' covered by the Access and Benefit Sharing (ABS) principle.

In 2016 a community seed bank was established in Naxi country, at Stone Village, Yunnan province. Three years later, UNEP and the Institute for Geographic Sciences and Natural Resources Research of the Chinese Academy of Science hosted a vibrant international conference in Beijing on 'Women, Biodiversity and Sustainable Food Systems in a Changing Climate'. Acknowledging the fact that out of 260 million smallholders, women make up 60 per cent of the country's agricultural workforce, the meeting target was 'multiple stakeholders' from farmers to academics, government ministries, multilateral agencies, NGOs including the Gates Foundation, Biodiversity International and Oxfam. The UN rationale was: 'A broader socio-ecosystem analysis and inclusive approach for sustainable food systems is proposed for sex/gender sensitive and supportive policies and actions for women's empowerment and agrobiodiversity enhancement.'[6]

While the various local seed banks remain to be integrated with China's national gene bank system, the project is described as a design innovation fostering heritage as well as value-adding through genetic improvement of crop landraces and wild medicinal plants. It is envisaged that enterprise, partnership, marketing and collaboration will be encouraged by using the

> internationally accepted system of access and benefit-sharing of genetic resources, as regulated by the International Treaty on Plant Genetic Resources for Food and Agriculture (ITPGRFA) and the Nagoya Protocol on Access to Genetic Resources and the Fair and Equitable Sharing of Benefits Arising from their Utilization to the Convention on Biological Diversity.[7]

So there are two kinds of local farming initiative in China right now. The first approach is socially transformative, emphasizing self-sufficient production, distribution and consumption, hand in hand with cultural autonomy and epistemic justice. The second approach frames seed sovereignty in terms of national and international market objectives. In this, scientists, governments and UN policymakers assist capitalist enterprises in accessing germ plasm to genetically engineer new products for sale. Activists at the 1992 Rio Earth Summit called this process 'biopiracy' wherein UN policy enables business, in this case Big Pharma, to freely extract and patent genetic material developed over many years by indigenous labour and knowledge.

Perhaps the key distinction is found in the contrasting terms 'food sovereignty', which implies self-determination for subsistence needs, whereas 'food security' implies business as usual. The High Level Panel of Experts on Food Security and Nutrition (HLPE) favours agroecology because of its closed cycles and participatory knowledge making. It also has advantages for climate, water, biodiversity and, indeed, the UN Sustainable Development Goals. Even so, questions have been raised with the UN Food and Agriculture Organization over an announcement that it will strengthen collaboration with the trade association CropLife International and other advocates of the dominant industrial food regime. For instance, climate-smart agriculture (CSA) and sustainable agriculture intensification (SI) can involve

> an eclectic mix of inputs and processes such as herbicide-tolerant crops, toxic insecticides, genetically modified seeds and livestock, proprietary technologies and patents on seeds, energy-intensive livestock factory farming, large-scale industrial monocultures, big data and digital-based precision farming, carbon offset schemes, and biofuel plantations.[8]

Given the power imbalance between local communities and even nation states versus the international corporate sector, the risk is that local farming will become disembedded from its social context and co-opted by the global monoculture. It is to be hoped that in China, the government will shield its unique agricultural heritage from such predations as these.

Whose Capacity?

It bothers me that the G8 and 'development experts' always talk about 'capacity building' for non-industrial communities in the Global South, because as I see it, capacity building is just what's needed in the Global North! And I'm not alone in this. In Anchorage 2009, the Indigenous Peoples' Global Summit pointed to the ecological debt notched up by affluent societies as main contributors to global warming. They spelled out their own vision of self-managed local economies based on food and energy sovereignty. And they told the UN that the time had come to hold Technical Briefings by indigenous peoples on traditional knowledge and climate change.⁹ Governments and international agencies have not given due credit to the capacities of peoples at the margins of capital. There's a tacit environmental racism too, in letting these others pick up the tab – like when a nuclear waste dump is put in Aboriginal country or when a UNFCC scheme like Reducing Emissions from Deforestation and Degradation (REDD) converts the livelihood of Kalimantan farmers into a carbon sink for our coal-based consumer lifestyle.¹⁰

The rhetoric of international 'partnership' does little to soothe the assault. True, conservation NGOs are anxious about the impact of rising seas on island peoples, but racism appears again in the claim that the populations of India and China are the biggest threat to global warming. For ecological footprint studies show that consumption *per capita* in China is negligible, compared with the average individual footprint in Australia or the United States.¹¹ The ecological footprint indicator is a terrific instrument for keeping policymakers honest, but mathematical formulae will not shift the totally ethnocentric idea that the ecomodernist model of production is the only viable way. In fact, the fixation on adjusting input/output parameters simply delays a more thoughtful response to climate change. This crisis asks us to stand back and look at why the humanity-nature metabolism has been configured so badly. Metabolism is the process by which humans take from nature, digest and give back in return – and cultures across the world have devised different ways of managing it. The pioneering ecological economist Nicholas Georgescu-Roegen made it the centrepiece of his new discipline, bringing biological systems and thermodynamic principles into economic reasoning, although still today, sustainability science deals largely with the tip of the production iceberg.¹² Most transfers between humans and nature are meta-industrial – non-monetized – and not even named by economists.¹³

Sociologists give various explanations for this strange modern disconnect from 'the natural'. Recall that the ecoMarxist analysis of John Bellamy Foster attributes it to the metabolic rift between town and country, with corporate globalization and free trade now exporting this rift across the Earth. Peter Dickens, also a socialist, sees an alienative consciousness and ecological abuse as an inevitable outcome of the industrial division of labour. And the more technologically 'improved' daily life is, the more people lose a feeling for their own organic embodiment as nature. Feminist Silvia Federici uses the word 'amnesia' to describe this psychological splitting.[14] The patriarchal-colonial-capitalist Humanity/Nature divide is very clear in the language of ecological economics: 'Natural capital can be considered the planetary endowment of scarce matter and energy, along with the complex and biologically diverse ecosystems that provide goods and services directly to human communities . . . [water recycling, pollination, and so on].'[15] Here, 'scarcity' is treated as an ontological constant, rather than a man-made anomaly; living systems are pulverized for turning into profitable goods. Metabolic flows are pulled apart and abstracted as 'factors'. There's little grasp of human co-evolution with the environment. And what's more, the sex/gendered, racialized and class origins of economics itself are bypassed.

Of course, there are also innovative moves in sustainability science. Herman Daly – albeit from the World Bank – matches the canon of efficiency with a call for global justice. He knows too that biological time is slower than economic time and that intergenerational equity calls for thinking with a long horizon.[16] Yet the heritage of neoclassical economics remains active in Daly's work. His core variables of scale, distribution and allocation function in an *ad hoc* system, whose *1/0 imaginary* boundaries are never justified. His cybernetic analogies reify the market much as the old hidden hand of liberalism did. And descriptions of economic processes in the passive voice set up a sense of inevitability that deflects people from making change. As a sociologist of knowledge, I am only too aware that the transformative potential of any discipline is stymied as long as cultural bias in its analytic tools goes unnoticed. Thus, Daly and confreres omit to ask: Who is it that decides on scale? Who distributes to whom? Who is entitled to make allocations? And, Why? Sure . . . conferences in ecological economics now include sections on peasant and indigenous societies and sometimes even host a feminist symposium. But my impression is that these are add-ons – 'problem areas', 'distribution conflicts' or 'externalities' waiting to be assimilated to the master map.

Meta-Industrial Labour

If the androcentric amnesia of industrialization were shaken off, would it then become respectable to talk about other ways of satisfying human needs? For example, a 2007 Food and Agriculture Organization report noted that the greater part of the world's food supply is actually produced by peasant cultivation.[17] The celebrated ecofeminist Vandana Shiva tells how this eco-sufficiency is achieved in northern India. It is women who manage the integrity of ecological and human cycles. 'As healers, they gather medicinal herbs among the trees; as catalysts of fertility, they transfer animal waste to crops, returning by-products to animals as fodder.' Their daily round – protecting natural sustainability and human sustenance – is an exemplar of scientific complexity in action.[18] First Nation peoples in Australia too make the seasonal walk through country with deliberation and disciplined harvesting to ensure renewal. Three hours work a day suffices in this bioregional economy. According to John Gowdy, a student of Roegen, the nomadic hunter-gatherer rarely extracts more matter or energy than needed for maintenance.[19] Don't get me wrong, I am not saying that everyone should head for the hills but arguing that the epistemology of these alternative production models provides essential capacity building for a global regime like ours, staring at a 'wrong way go back' sign. In regions where communal land is undisturbed by European ideas of development, self-managed indigenous economies are both sovereign and synergistic.

Let's face it: the economy of permanent consumption and 'green conversions' is not only ethnocentric, it also fails the thermodynamic test. Ideally, a bioregional socio-economy will meet livelihood needs as far as possible within its natural catchment. In his call to *Dwellers in the Land*, Kirk Sale proposed the following prerequisites for this:

> to know the Earth fully and honestly, the crucial and perhaps only and all encompassing task is to understand *place*; the immediate specific place where we live. The kinds of soils and rocks under our feet; the source of the waters we drink; the meaning of the different kinds of winds; the common insects,

birds, mammals, plants and trees; the particular cycles and seasons, the times to plant and harvest and forage . . . the limits of its resources; the carrying capacity of its lands and waters; the places where it must not be stressed.[20]

Most peasant and indigenous provisioning exemplifies this vernacular sustainability science. So we find:

- The consumption footprint is small because local resources are used and monitored daily with care.
- Closed-loop production is the norm.
- Scale is intimate, maximizing responsiveness to matter-energy transfers in nature, so avoiding entropy.
- Judgements are built up by trial and error, using a cradle to grave assessment of ecosystem health.
- Meta-industrial labour is intrinsically precautionary, because it is situated in an intergenerational time frame.
- Lines of responsibility are transparent – unlike the buck passing that mars bureaucratized economies.
- With social organization less convoluted than in urban centres, synergistic problem-solving can be achieved.
- In farm settings and in wild habitat, multicriteria decision-making is simply common sense.
- Regenerative work reconciles time scales across species and readily adapts to disturbances in nature.[21]
- This economic rationality distinguishes between stocks and flows. No more is taken than is needed.
- It is an empowering work process, without a division between workers' mental and manual skills.
- The labour product is enjoyed or shared whereas the industrial worker has no control over his or her creativity.

- Such provisioning is eco-sufficient because it does not externalize costs on to Others as debt.
- Autonomous local economies imply food and energy sovereignty.[22]

These principles reveal a peoples' science that vies closely with the advice of good environmental consultants in the Global North. The trouble is that in growth economies, the advice is often shelved by governments under pressure from business; or, unwieldy administrative management foils the translation of principles into action on the ground. Beyond this, capitalist states are so dependent on resources from the meta-industrial periphery that existing sustainability practices of the global majority must remain unspoken.[23] However, technical briefings by indigenous peoples are indispensable if international decision-making is to become coherent and democratic.

Agri-tech

In agriculture today, there is undue international influence by the World Economic Forum and Big Pharma. Dow, Dupont and Monsanto-Bayer Crop Science own 50 per cent of the patented seed industry and 70 per cent of agrochemicals. Meanwhile, Big Data firms want to 'rationalize' agricultural practices by digitalization, and farm machinery manufacturers and brokers are another part of the mix. Since 2021 India has been racked by massive Farmers' Strikes in response to government efforts to industrialize the sector. Three years later, farmers in the Netherlands, the United Kingdom and, in fact, right across Europe are on the streets too, tractors and all, resisting government proposals for 'agri-tech'. That is to say 5G towers in fields to run robot farmers, replacement of natural seed and cattle by GMOs, with insect biomass for manufactured protein. Activists in Somerset note that this scheme already operates in parts of the United States, which may help explain why philanthropist Bill Gates is the country's biggest landholder. Small food producers across the world boycotted the 2021 UN Food Systems Summit, to organize a meeting of their own. The call-up list is extensive: People's

Coalition on Food Sovereignty (PCFS), PAN Asia Pacific (PANAP), Asian Peasant Coalition (APC), Arab Group for the Protection of Nature (APN), Arab People for Food Sovereignty (ANFS), Eastern and Southern Africa Small-scale Farmers Forum (ESAFF), Indigenous Peoples' Movement for Self-Determination and Liberation (IPMSDL), Coalition of Agricultural Workers International (CAWI), Asian Rural Women's Coalition (ARWC), Global Forest Coalition (GFC), People Over Profit (POP), Asia Pacific Research Network (APRN), IBON International, Asia Pacific Forum on Women, Land and Development (APWLD), Stop Golden Rice Network (SGRN), PAN North America (PANNA), A Growing Culture, Youth for Food Sovereignty (YFS). As an appeal to UN Secretary-General Antonio Guterres pointed out, recent UN policy is a serious violation of Article 10.1 of the UN Declaration on the Rights of Peasants and Other People Working in Rural Areas.[24]

As things stand, under the global corporate-designed system, 800 million people across the world go hungry every day, while ecomodernist food production emits some 40 per cent of global warming carbon emissions. How to apply the fourteen principles of meta-industrial provisioning in political decision-making? If democracy is to be protected, in a world increasingly paved over by the vanities of Eurocentrism, then vernacular science and 'subsidiarity' is the word.

11

Another Future Is Possible!

Holding Ground

Since the shock of the 2008 financial crisis, 'green finance' has provided a survival strategy for capital. This became very clear at Rio+20, anniversary of the first Earth Summit in Brazil. While the conference negotiating text was called *The Future We Want*, the big question, of course, was, 'Who is we?' For the organizing network Business Action for Sustainable Development (BASD), this 'we' was definitely 'the private sector'. As their Zero Draft put it:

> Multitudes around the world are part of the private sector, whether self-employed, entrepreneurs, farmers or small and medium sized as well as large multi-national enterprises. The private sector generates most of the goods and services that are utilised every day and therefore must be actively engaged to address the implementation gaps that have limited the achievements of sustainable development goals.[1]

For sure the private sector should be politically engaged, responsible as it is for the global ecological crisis, but this official portrayal of the private sector is falsely homogenizing. It invisibilizes several economic groupings, including workers – when it is actually peasants, mothers, fishers and gatherers who meet basic needs for the majority of people around the world.[2] More than this, the meta-industrial class already provisions in ways that meet the goals of precaution and sustainability.[3] Following in the footsteps of the Business

Council for Sustainable Development (BCSD), which had steered the United Nations at Rio 1992, now it was the BASD that promoted the corporate sector as sponsor and ideas man for reframing international governance. Meanwhile, the International Monetary Fund (IMF) announced a Green Economy Initiative (GEI) building on 'the strengths of the market-based economy' but supported by a more 'coherent institutional framework'.[4]

Andro Others

It was no surprise that the international peasant organization *Via Campesina* read the Rio+20 Green Economy as yet another phase of 'green structural adjustment programs', seeking 'to align and re-order the national markets and regulations to submit to the fast incoming "green capitalism"'.[5] By the turn of the millennium, workers, indigenous, womens and ecological voices were joining up. People with life-affirming skills and values were active in peasant food sovereignty and indigenous environment networks; women were especially active as anti-toxics campaigners and peace activists. Faced with the new Rio+20 event, the WSF Thematic Social Forum now circulated a powerful synthesis of shared concerns under the head *Another Future Is Possible: Come to Re-Invent the World at Rio+20*.[6] Each social movement had its own political objectives and discourse, which WSF advocates still strive to balance; but the emergence of Occupy in the United States and Indignados in Spain stirred a new round of Left self-examination. Beyond this, the Canada-based action group Erosion, Technology, Convergence (ETC) injected a strong critique of biocolonialism into the peoples' WSF agenda.

Capital accumulation through the proposed Green Economy would rely on rapid technological innovation, but that means more resource extraction, biodiversity loss and energy pollution. In the words of the ETC organization, a long-time Science for People advocate:

> The big idea is to replace the extraction of petroleum with the exploitation of biomass (food and fibre crops, grasses, forest residues, plant oils, algae, etc.). Proponents envision *a post-petroleum future where industrial production (of plastics, chemicals, fuels, drugs, energy, etc.) depends – not on*

fossil fuels – but on biological feedstocks transformed through high technology bioengineering platforms. Many of the world's largest corporations and most powerful governments are touting the use of new [but untested] technologies including genomics, nanotechnology and synthetic biology to transform biomass into high-value products.[7]

Clearly, the Green Economy is a heavily Earth demanding strategy from an environmental perspective.

The key substantive issues for Rio+20 discussions were energy access and efficiency; food security and sustainable agriculture; green jobs and social inclusion; urbanization; water management; chemical wastes; oceans; risk and disaster amelioration. Optimistically, 'greening' the global capitalist system was deemed to be an 'integration of economics and ecology'. At the same time, business and UN agencies argued that 'innovative instruments' for financing this must be consistent with 'the Doha Development Round of multilateral trade negotiations'. The big-picture steps towards this contradictory hegemony were:

- Moves to transform UNEP into a World Environment Organization
- Moves to assess the feasibility of Earth System Governance
- Moves to explore a new Global Financial Architecture.

The leaders of the patriarchal-colonial-capitalist imperium – the World Bank, the International Monetary Fund, the regional development banks, UNCTAD and the World Trade Organization – were all asked to consider the ecosystemic implications of their decisions. And in so doing, the agents of social dislocation gained fresh political legitimation.

With guidance from UNEP, *The Future We Want*, also known as the Zero Draft, spelled out the neoliberal terms of reference and potential outcomes for Rio+20. It built on earlier agreements such as the Johannesburg Declaration, the Monterrey Consensus, the Doha Round, the Istanbul Programme for Least Developed Countries and the Bali Strategic Plan for Technology Support and Capacity Building. The Zero Draft also endorsed the 1992 Rio principle of 'common but differentiated responsibilities' as a way to define fair relations between Global North and South. But while the need to remove poverty from the planet was prominent in the negotiating text, other critical 'p' words were missing – like power, for example, and profit. Thus Rio+20 came in spinning with networks, promo agencies, think tanks, websites and conferences. Canada's International Institute for Sustainable Development (IISD) offered itself as a comprehensive 'knowledge management project' in preparation for the pending 'innovation culture'. Proposals arrived online from serious bodies like the New Economics Foundation and the World Future Council. The feminist network Women in Europe for a Common Future was engaged; ministers from the Congo called for a new inter-Governmental architecture; and global policy meetings were conducted by facilitators with buzzy names like Bright Green Learning. In London, A Planet Under Pressure gathering was held under the auspices of the Royal Society. Described as giving scientific leadership to Rio+20, plenary lectures were offered by the World Bank, a Shell Oil Company vice president, and the United Kingdom chief scientist. Attendance registration was set at GBP 400 per head. Elsewhere, Lund University in Sweden, the Australian National University and UN University in Tokyo were being funded to host conferences on Earth System Governance.

But thinking publics would be paralysed by a maze of acronyms such as IEG (International Environmental Governance); 10YFP (10-Year Framework of Programmes on Sustainable Consumption and Production); CPR (Committee of Permanent Representatives); EMG (Environment Management Group); IPBES (Inter-Governmental Science-Policy Platform on Biodiversity and Ecosystem Services); GLISPA (Global Islands Partnership); ISFD (Institutional Framework for Sustainable Development); SAICM (Strategic Approach to International Chemicals Management); UNON (United Nations Office at Nairobi); and even COW (Committee of the Whole)! Could political language be more mystifying, exclusionary and disempowering? Rio+20 forged a new discourse of international governance, a shared set of social and material expectations across nations, classes and bodies. But market logic, like carbon trading, geoengineering or climate smart agriculture, does not restore broken life-support systems in nature. The Worldwatch Institute calculates that 60 per cent of 'Nature's services' have been destroyed by industrialization since the Second World War.[8] The trouble is that economics is about abstract constructs, whereas ecology is about matter-energy flows. There is a profound cognitive disjunction between the two disciplinary lenses. International consensus on a totalization like the Green Economy is that it can do little for sustainability – or democracy. Rather, what this concept designed by free traders does is reinforce the unequal patriarchal-colonial-capitalist exchange as already exists. In fact, the Rio+20 GEI simply introduced the next stage in a centuries-old history of Eurocentric colonization.

Major Groups

The *Androcene* system of upwards capital accumulation only functions on the back of a surplus provided by 'naturalized' Others. Thus, as noted, capitalism is built on a social debt to exploited workers; an embodied debt to unpaid women for their reproductive labour; a livelihood debt to peasants and indigenous peoples for appropriating their land; and a generational debt to youth.[9] So too, history has shown that this extraction from the human peripheries of capital relies on the development of a comprador class, groomed with incentives by the colonizer. This 'cloning' is the real meaning of 'development', and such power relations are enacted today through the UN machine, through the business world and even through universities. High-level consultations for Rio+20 succeeded in taming not only well-paid bureaucrats but scientists and academics too. Sometimes, the most enthusiastic intermediaries for the capitalist class come from marginalized populations. Women especially can be vulnerable to the enticements of comprador status as they try to obtain better conditions for their communities. At Rio+20 the World March of Women and various feminist groups reading the Zero Draft took on this political double bind:

> We call for removing barriers that have prevented women from being full participants in the economy and unlocking their potential as drivers of sustainable development, and agree to prioritize measures to promote gender equality in all spheres of our societies, including education, employment, ownership of resources, access to justice, political representation, institutional decision-making, care giving and household and community management.[10]

While sex/gender mainstreaming seems benign, the criterion for equality under the *Androcene* is 'the masculine universal' – an idealized image of the emancipated woman as one who can live like a middle-class, white man. Women's inconvenient sexual embodiment will often be neutralized with technological help. In this way, unique insights learned from reproductive labours are 'politically contained' at the elite level. Nevertheless, at the 56th session of the UN Commission on the Status of Women (UNCSW), UN Deputy Secretary-General Asha-Rose Migiro highlighted the fact that rural women constitute a quarter of the world's population; grow the majority of the world's food; and perform most unpaid care work. There is no doubt that this situation merits recognition but not when support for water infrastructure and renewable energy benefits business donors more than recipients. Likewise, when UN-Women's executive director Michelle Bachelet pointed to barriers in women's empowerment and called for more sex/gender sensitivity in national budgets and business, the effect was to enhance liberal assimilation. Like micro-credit schemes, such measures quietly recruit women as players in the capitalist system.[11]

As with the originary 1972 Stockholm Conference on the Human Environment (UNCED), the motor of Rio+20 would be UNEP, now under Executive Director Achim Steiner. However, at the pinnacle of the Rio+20 hierarchy was Secretary-General Sha Zukang, a Chinese career diplomat. He was less upfront than Maurice Strong, the Canadian businessman who had brokered Earth Summit 1992 but very active on the Sustainable Energy for All Initiative. The vision was that after Rio+20, the UN Commission on Sustainable Development (SD) should be upgraded to Council status – CSD becoming SDC – and the United Nations Economic and Social Council (ECOSOC) would have a strong outreach role. In preparatory dialogues for Rio+20, the term 'civil society' was used often. However, this essentializes social differences and dissolves a myriad of grassroots struggles under the bland formula of citizenship. Worldwide, people see national parliaments swept into the revolving door of suits and find they have nowhere to go but the streets. This is why the grassroots network World Social Forum was born a few years after the confrontation with multilateral agencies at Seattle. It is why the US Occupy Movement for direct democracy broke out following the second financial crash in 2011. And it is why at Rio+20, just like Rio 1992 before it, the Peoples' Summit would be located for security reasons a good distance from official UN and government proceedings. Nevertheless, the Peoples' Summit for Social and Environmental Justice in Defence of the Commons would dual power the UN Rio+20 from 15 to 23 June 2012.

As UNEP's *One Planet* magazine explained, getting the Rio event up means orchestrating three kinds of human – Inter-Governmental, Governmental and Non-governmental. In the first two sectors, state ministers or their stand-ins meet under the rubric of GCSS-12 / GMEF – that is to say, the UNEP Governing Council/Global Ministerial Environment Forum (Special Session 12). These national representatives are deployed to spell out a mix of new Green Economy models 'tailored to different local and national conditions' – at once 'pro-growth' but based on 'a measurement of well-being' that goes beyond GDP. Governments are asked to configure the Millennium Development Goals (MDGs) into their policy, albeit heavily criticized by activists.[12]

At Rio+20 the non-governmental sector was marshalled under the UN acronym GMGSF-13, standing for Global Major Groups and Stakeholders Forum in its 13th Session. Led by Felix Dodds, a designated space was made here for Women's groups, Children and Youth, Indigenous Peoples as well as NGOs, Labour and Unions, Business and Industry, the Science and Technology community and Local Authorities. There was also scope, possibly *ad hoc*, for regional opinion makers. But with no acknowledgement of the dynamics of power and profit as economic levers of capitalism, there was a good deal of sociological fudging in these Rio+20 consultations. Preliminary arrangements gave a nod to rights-based 'vulnerabilities', such as sex/gender and ethnicity, but class analysis was absent. Instead, business delegates followed the UK's New Economics Foundation in talking about 'joining the dots' of 'the three pillars' social, economic and environmental. A skilled transdisciplinary analysis would be required to tease out the complexities and contradictions inside such a crude functionalist agenda.[13] UNEP believed that Major Groups and Stakeholders would readily converge on the global Green Economy theme. Meanwhile, the corporate sector was urged by UN Secretary-General Ban Ki Moon to sign on to a Global Compact, circulating as a rights-based credo of ten principles. The International Trade Union Confederation (ITUC) also came on board with the Green Economy approach and the idea of a new architecture of global governance. In addition, ITUC prioritiszd access, right principles, concrete targets and accountability. So too the International Council for Science (ICS) joined seeking clear definitions and measurable implementation. But the danger was that if concerns of Major Groups got too tied up with operational matters, then capitulation to the *status quo* could happen by default.

Left proposals for Rio were floated on a citizen audit of global debt, alternative currencies, minimum and maximum wage levels, bank socialization, trade regulation, a financial transactions tax and an end to land, water and biodiversity piracy. Some may see such campaigns as reformist or transitional, but they are educational, if nothing else. Significantly, the only Major Groups to express 'material alternatives' were people whose work is to reproduce natural processes. Thus,

- Women wanted their unpaid domestic contribution valued.
- Indigenous peoples wanted secure land rights.
- Peasant farmers wanted attention to local food sovereignty.

The local economic provisioning and caregiving of these meta-industrial workers already exemplifies an alternative politics of social commoning and sustainability. Inhabiting the domestic and geographic peripheries of capital, the class of meta-industrial labour is still largely ignored, even by many on the Left. Yet in an era of environmental crisis, the very notion of such a class is powerfully integrative. It broadens the classic socialist preoccupation with productivist industrial workers. It transcends the divisive idealism of liberal identity politics such as 'feminism' or 'indigeneity'. Meta-industrial work is materially grounded in the reproduction of embodied and ecological processes. In maintaining the humanity-nature metabolism, this activity is transcultural, and it is not necessarily sex/gendered.

Meta-industrial labour already models the 'green jobs' that the UN, private sector and unions hope to 'generate' out of technological innovation. This unspoken class already meets the needs of millions of people without destroying ecological cycles. Keynotes and committee chairs at Rio+20 should have been drawn from this class. By democratic counting they are the global majority, and by sustainability counting they are the skilled managers of 'Nature's services'. Where were these folk in the Green Economy Coalition (GEC) forming around UNEP? Instead, GEC associates led by Oliver Greenfield comprised the following constituencies: *Vitae Civilis*, Consumers International, International Institute for Environment and Development, World Wildlife Fund, Biomimicry Institute, International Trade Union Confederation, the International Institute for Sustainable Development, International Union for the Conservation of Nature, Ecologic Institute, the Bellagio Forum for Sustainable Development, Aldersgate Group, Philips Global, Development Alternatives, the International Labor Organisation, SEED Initiative, World Business Council for Sustainable Development, Global Footprint Network, Ethical Markets Media, the Caribbean Natural Resources Institute, The Natural Step and Eco Union. In truth, even if meta-industrials had scored a seat at this table, their voices would have gone unheard in the din of technocratic discourse.

The Green Economy Coalition described its mission as 'resilience within the limits of the planet', but from an alternative perspective, it was highly Eurocentric and masculinist in the conception of research and product design, 'partnerships for local entrepreneurship', grant giving, educational forums and reporting. The programme entertained a mixed bag of themes: Millennium Development Goals (MDGs); equity yet inclusive governance; market reform yet competition; green jobs yet finance for technology; workplace standards yet 'best practice'; and transitioning. There was interest in the idea of an International Environment Court, although the question of global power relations was not laid out on the table. At a G20 meeting in Mexico late February 2012, President Felipe Calderon intimated that by funding technology transfer the Global North could compensate for both climate change and the ecological debt of colonization. But the environmental imposts of industrial technology were passed over in favour of opportunities for growth: 'current high energy prices open policy space for economic incentives to renewables . . . investors are looking for alternatives given the low interest rates in developed countries, a factor that presents an opportunity for green economy projects.' The G20 communiqué rallied by asking the Organisation for Economic Co-operation and Development (OECD), the World Bank and the UN to prepare a report 'inserting green growth and sustainable development policies into structural reform agendas, tailored to specific country conditions and level of development'.[14]

Coloniality

The Green Economy notion is contradictory for at least two reasons. First, it is embedded in capitalism, a system whose *raison d'être* of profit depends on extractivism and cost transfer. These colonial externalities are experienced by Others in the structural violences of social debt, livelihood debt, embodied debt and intergenerational debt. Second, the official Green Economy is embedded in industrialization, and notwithstanding the rhetorics of dematerialization, even ecologically modernizing digitalized production cannot avoid energy and resource draw downs. In the human metabolism with nature, industrial technology never solves a problem; the best it can do is displace it. The displacement may be spatial – shifted on to less powerful social sectors – or displacement may be temporal – shifted on to future generations. As capitalism exhausts the planet's capacity to provide material throughput for industrial value-adding, it is not just high-tech 'renewables' but new global institutional architectures that push against natural limits by enacting constitutional powers for Earth System Governance. Even so, around 150 members of the official Rio+20 constituency – including Malaysia, Congo and Peru – were hoping to see UNEP transform into a specialized agency with powers like the WTO and capacity to simplify the 900 odd Multilateral Environment Agreements (MEA). Some members favoured social change strategies based on global treaties or regional conventions or deals. The Zero Draft confirmed the neoliberal character of *The Future We Want*.

The Green Economy balances precariously on circuits of 'financial capital, human capital, natural capital, and physical capital'. By imputing 'economic value' to the life-giving capacities of 'Nature's services' so-called metabolic flows must be reduced to tradeable units. This reductive fiction is an epistemological violence. In June 2012, young people, small farmers, workers, squatters, grandmothers and indigenous gatherers converged on Rio+20 to oppose the deadly commodification of life. But environmentalists like the Global Footprint Network give up the game when they say that 'billions of dollars of investment' will be necessary to make sustainability real. Again, Janez Potočnik, European Commissioner for Environment, invited funds from non-traditional sources to 'green' the Global South. Edna Molewa, South Africa's Minister for Water and Environmental Affairs, hoped to see public-private partnerships multiply. As the global private sector weakens governments, leading to cash-starved public universities, these are co-opted with business donations for centres of excellence. Advocacy networks like Environmental Justice Organizations, Liabilities and Trade (EJOLT) walk such a tight rope. Sponsored by European Commission research moneys, EJOLT runs an important website, database, workshops, policy papers, focusing on the fair 'distribution' of 'development benefits'. However, such projects have to take care not to absorb grassroots energies, on the one hand, or allow governance institutions to exercise 'repressive tolerance', on the other.[15] Middle-class activists need to think hard about tactics.

On the US West Coast, wired-up activists turned to social media to find 'a fast-mutating array of high-tech opportunities for creating solutions to social and economic problems'.[16] But IT, as such, is part of the problem – being both a voracious energy user and toxic to the bodies of Chinese and Mexican assembly line workers. Few alternative globalization activists have examined cradle to grave externalities of the digital revolution. As Business Action for Sustainable Development put it:

> the corporate vision is thoroughly framed by the technological *a priori*: Collaboration and collective action on innovation and technology development and their appropriate deployment via sustainable consumption and production (SCP) *are at the heart* of greening economies.[17]

Here, even Sustainable Consumption and Production (SCP) is scientized and legitimized with an acronym – and it is telling in an era of false needs that consumption precedes production in this slogan. But 'at the heart' of this hyper-industrialized patriarchal-colonial-capitalist system, who collaborates with whom? Capital accumulation in the great world cities has always relied on forcing food-sufficient peoples off their land to become factory workers and consumers. And today, in order to keep the SCP model afloat in the Global North, women remittance migrants from the Global South are sent across the world as cheap domestic labour abroad and foreign exchange for governments at home. Meanwhile, child-caring grandmothers pick up the transferred cost of this reproductive labour supply chain. In short: the 'knowledge base' enthusiastically propagated by UNEP and IISD – arguments for the global Green Economy are seriously lacking, not just in ecological literacy but in sociological literacy as well.

In the 1970s, the activist slogan was 'live simply so others may simply live'. But this commitment was overtaken in the 1980s–90s by the rise of 'professional environmentalism' championed by business through the UN's various sustainable development programmes. By the new millennium, Latin American peoples were revitalizing Left politics, with Earth-centred Constitutions in Ecuador and Bolivia that recognized the Rights of Mother Nature. So too, the history-making 2010 Cochabamba Climate Summit, hosted by the Women and Indigenous Peoples of Bolivia, advanced the principle of *buen vivir* or *sumak kawsay*. The precise meaning of these words is unique to Andean cultures but versions of 'living well' are adopted everywhere now by commoning activists. Broadly speaking, wherever resources remain free of capitalist appropriation, 'the dots' are still joined, and people – eco-sufficient meta-industrial communities – enjoy autonomous ecologically sensitive provisioning. The classic Eurocentric hierarchy of Man over Nature, and the metabolic breakdown that results from it, is unknown. At the margins of the colonial imperium the Earth is valued for itself, not simply as a resource for human profit. And where economics is an embodied materialist practice, people rediscover identification and belonging in working together with nature. The call for this deep green future did not find its way into the Zero Draft.

A Biocivilization

In anticipation of Rio+20, the global grassroots network called World Social Forum (WSF) issued a working paper entitled *Another Future Is Possible: Come to Re-Invent the World at Rio+20*. Its chapters were infused with the vision of a 'biocivilization' – as distinct from the corporate managerial push into biocolonialism. The heads were thorough:

- Ethical and Philosophical Foundations: Subjectivity, Domination, and Emancipation

- Foundations for a Biocivilization – Education in a World Crisis – Knowledge, Science, and Technology

- Human Rights, Peoples' Territories, and Defense of Mother Earth

- Right to Land and Territory – Territory and Native Peoples – Sustainable Cities

- For the Right to Water as a Common Good – Health Is a Universal Right, Not Source of Profit

- Production, Distribution, and Consumption: Access to Wealth, Common Goods, and Economies of Transition

- Finance and a Fair and Sustainable Solidarity Economy

- The Green Economy: A New Phase of Capitalist Expansion – Energy Transition Is Urgent and Possible

- Political Subjects, Architecture of Power, and Democracy

- The Commons: A Kaleidoscope of Social Practices for Another Possible World – Governance and the Architecture of Power.[18]

The document, backed by the World Social Forum Charter of Principles, tackled the hegemonic push for Earth System Governance head-on, but there were internal challenges. One proposal coming from WSF to Rio+20 was simply to 'close it down.' Other tensions appeared over strategy, a collision between the commoning style of horizontalists like Occupy and the more formal vertical organization style of the conventional Left.[19] While Occupy movement folks wanted to talk about hydroponics, the Left felt a need to give the prefigurative crowd a sense of how their grievances interlock with complex international forces. For example, unemployment is tied to offshore restructuring; the housing crisis is tied to an unregulated world financial system; poor health services are tied to excess military spending. But the characterization of WSF as 'old movement vertical' and Occupy as 'new movement horizontal' was a facile dualism and it made the movement vulnerable to external manipulation.[20] In fact, in everyday organizing, a mix of strategy differences should be useful. Thus, Stephen Lerner predicted:

> When the passion, fearlessness and vision of Occupy intersects with the resources and membership of community groups and unions, we'll find the sweet spot that makes it possible to force the richest to negotiate with the rest of us. It is where these two worlds meet – horizontal and vertical – united around common issues and enemies that we create the potential to start winning together.[21]

Communication with the wider world was the real challenge. As WSF activist Chico Whitaker pointed out, the movements barely add up to a global demographic of 1 per cent; and while many people struggle to survive, many more are content with their commodity comforts.[22]

Real Green Jobs

Remembering always that 'the economic bottom line' is really an ecological one, the WSF working text *Another Future Is Possible!* still inspires thoughtful transformative moves. Thus the official Rio+20 Zero Draft provoked the following interventions under the head 'Stop the Green Monster, the Future We Don't Want':

- *Celebrating* the Rights of Mother Nature in the Decade of Water for Life. Just as human bodies are joined to ecosystems, land is joined to water and water to air. Through living plants, evapotranspiration cycles fertilize land and cool the atmosphere. This ecological rationality is lost with market-oriented climate solutions like the Clean Development Mechanism (CDM) and Reducing Emissions from Deforestation and Degradation (REDD).
- *Endorsing* the proposal for A High Commissioner for Future Generations and insisting s/he focus on a protective treaty covering the health and social costs of new technologies.

- *Supporting* the UN Declaration of the Rights of Indigenous Peoples and matching it with a UN Declaration of the Rights of Peasants. Both should be cross-referenced to poverty, women, children, youth, land rights, livelihood and food sovereignty – that is, real 'green jobs'.
- *Converting* UNEP's 'mix of policy options' into a genuine exploration of non-violent decentralized alternative modes of provisioning: commoning and collective rights, cultural diversity, local autonomous gift economies from 'simply living' in the Global North to *'buen vivir'* in the South.
- *Seeking* agreement that governments and multilateral agency decision-making and institutional modelling uphold the distinctions between public and private and science and policy. The Inter-Governmental Panel on Climate Change (IPCC) is a good place to start applying this.
- *Challenging* the rhetorical 'integration of social, economic and environmental pillars' by demonstrating how the very machinery of a capitalist economy, the dynamics of power and profit, is inherently incompatible with social justice and ecological sustainability.
- *Demanding* satisfactory attendance in training courses on socio-ecological literacy as prerequisite to participation in Rio+20. Workshops will include exercises in political reflexivity to enable delegates from the Global North to recognize colonization in all its forms. Current training courses by the World Business Council for Sustainable Development are a travesty of education.[23]

Given the fundamental link between sustainability and peace, the WSF Thematic text spoke of opening up the Security Council to 'new actors' – states, regional organizations, global networks – with a new balance of power based on bodies appropriate to watching over life, peoples and planet. Engaging with the international agenda helped broaden the political horizon and ground the political subjectivity of both WSF and Occupy – just as it will do with new partisans like Extinction Rebellion today.[24]

12

Green New Deals

For Globalization Lite

Early in the twentieth century, colonization of the 'geographic periphery' was recognized by Rosa Luxemburg as the essential source of materials, labour and markets for capital accumulation.[1] Subsequently, other feminist socialists in the Global North would recognize a parallel exploitation of the 'domestic periphery' based on women's freely given household 'reproductive labour' time. A third level of subsumption is highlighted by the concept of 'ecological debt'. Could the old androcentric dualisms of Centre versus Periphery, Masculine versus Feminine, Humanity versus Nature be hiding a political alternative to capitalism? Not all areas of the globe are 'evenly developed' or integrated into the modern economy, and many people even strive to stay free of it. They understand well how ecologies and bodies are regenerated by the metabolic value of their labours.[2] Only this kind of reproductive economy is viable in the longer term. But there is little cultural analysis across the popular international Green New Deal (GND) programmes offered so far. In the ecomodernist mainstream, ecological crises are addressed in Keynesian terms as a failure of governments to manage markets. The call is for technological solutions, applied as 'green welfare,' in a context of social democratic renewal. The deconstructive reading that follows offers a lesson in how not to think about humanity-nature relations; as Kolinjivadi and Kothari say:

What is easily forgotten in eco-friendly' talk, is just how development models of the Global North are structurally founded on *dehumanisation*, in which hundreds of millions across the globe are seduced and stripped of their diverse ways of knowing the world, and dumbed down into passive onlookers and screen junkies, unable or unwilling to acknowledge (much less act upon) the consequences of their consumption patterns.[3]

So dear reader, if you have lost patience already with the contradictory policies of ecomodernism, skip to the end of this chapter and move on to the next! You can always come back if you want to look something up.

UK and UNEP

In 2008, *A Green New Deal: Joined Up Policies* was launched by the United Kingdom's New Economics Foundation (NEF).[4] It is a thoughtful proposition, which seeks to damp down the contemporary growth ethic. However, the deal is squarely framed by productivist economics, with its emphasis on banking and securities regulation, low interest rates, controlled lending, a Tobin tax on capital movements, minimizing tax evasion and debt cancellation instead of bailouts. Like the Stern Review, it encompasses a managerial agenda of energy audits via renewables, technological efficiency, retrofits, forest protection and zero waste. The brief also considers social lifestyle and living density, community building, local economies and food miles – the latter exposing the cost of transport and refrigeration in global supply chains. After all, in the words of author Tim Jackson of the UK Sustainable Development Commission, 'prosperity consists in our ability to flourish as human beings – within the ecological limits of a finite planet . . . above a certain point – around US$15,000 a head, more GDP growth stops delivering more happiness.'[5]

In the same year, 2008, the Division of Technology, Industry, and Economics of the United Nations Environment Programme (UNEP) brought out its *Global Green New Deal*. The press release read: 'Green New Economy Initiative to Get the Global Markets Back to Work'.[6] Designed as a tool kit for governments, it develops earlier work from the G8 study group for the Economics of Ecosystems and Biodiversity (TEEB), the ILO, the International

Trade Union Confederation and the International Organization of Employers. It is written with assistance from the European Commission, Deutsche Bank and the World Bank's Global Environment Facility, all with an eye to the G8 Summit. The stated goals of the *Global Green New Deal* are: valuing and mainstreaming Nature's services into international accounts; generating employment through green jobs; developing policies and instruments for the economic transition. The initiative prioritizes: clean energy, clean technologies and recycling; rural energy, renewables and biomass; sustainable and organic agriculture; ecosystem infrastructure; Reduced Emissions from Deforestation and Forest Degradation (REDD); sustainable cities; green building and transport.

This is certainly comprehensive in scope, albeit a version of Mother Nature that is still hinged to the market. In fact, the speculative hypereconomy is offered as a further new deal option by bundling together: 'US weather derivatives and other insurance linked products to make them attractive to investors.' UNEP is brimming with success stories. It notes that already nations in Africa, Asia, the Middle East and South America have set renewable-energy targets; that in China, 600,000 people are employed in the solar-thermal industry; and in India over 100,000 homes are equipped with solar power. The Clean Development Mechanism (CDM) is assisting a hydroelectricity programme for Madagascar and energy generation from sugar cane waste in Kenya. The document talks about 'securing livelihoods' and goes some way towards recognizing the undesirability of differential benefits by class. Differential economic benefits by sex/gender are not registered.

The UNEP *Global Green New Deal* is rather more environmentally grounded than many similar propositions, and this reflects its international framing, with attention to rural economies and natural habitat in the Global South. It claims that 40 per cent of the global workforce are farmers and it remarks on the highly destructive impact of agricultural subsidies – amounting to some US$300 billion around the world annually. As FAO reports will confirm, an irrefutable body of research shows that organic agriculture and integrated pest management are not only more resistant to climate stress but improve soil fertility, biodiversity, water control, carbon sequestration and crop yields. Additional research sources indicate that organic farming could actually feed the current world population and even a larger one.⁷ The benefit is doubled where perennial crops are used. Farmers not only receive higher prices for organic produce, especially after certification, but also income is saved by not needing to buy fertilzer, pesticides or GM seed. In terms of social benefits, organic production is knowledge intensive and enhances community bonding. Even more significant is the fact that the majority of world food producers are women.

The UNEP brief calculates that deforestation is currently responsible for 20 per cent of greenhouse emissions, and it expects that unless there is immediate intervention, by 2050 the accumulated loss of reefs, wetlands and forests will be equivalent to an area the size of the Australian continent. It recommends protection for endangered species by 'smart instruments' like 'cap and trade'. It supports marine protection – and points out that reefs provide value in fisheries, tourism and flood protection. Wetland deterioration is to be mitigated by biobanking schemes as devised in the state of New South Wales. Also in Australia, the Minerals Council has been found to support biobanking and profits from the resale of permits.⁸ In the view of a somewhat unselfconscious UNEP: 100,000 National Parks and protected areas across the world generate wealth via nature-based goods and services equal to around US$5 trillion but they only employ 1.5 million people.

UNEP puts the 'service value of Nature' at a trillion dollars higher than profits generated by the international automobile industry – although it is not clear how this figure is arrived at. In Mexico and Brazil, thousands of people are now paid to manage watersheds. If Nature is 'natural capital', UNEP notes 'the flip side of the coin' will be the massive benefits to be had from 'the green technological revolution' and the 'huge untapped job potential' of managing 'nature-based assets'. Conservation, therefore, yields use value, exchange value and what ecofeminists call metabolic value.⁹ However, rather than explore the potential for a society-nature metabolism based on the logic of reproduction, a thoroughly androcentric productivist model enforces the subsumption of material nature here. In fact, UNEP envisages that the global market for securing environmental products and services can double by 2020. In the words of Executive Director Achim Steiner: 'natural "utilities" that for a fraction of the cost of machines, store water and carbon, stabilise soils; sustain indigenous and rural livelihoods and harbor genetic resources to the value of trillions of dollars a year'.[10] At least the *Global Green New Deal* recognizes the need to avoid impacts on low-income, ethnic and First Nation groupings. For too often, economic rationalization means the enclosure of indigenous lands and creation of refugees by absorption of self-sufficient local livelihoods and autonomous cultures by global capitalism.

Transatlanic

Another GND version is the *Transatlantic Green New Deal* prepared by Worldwatch Institute for the Boell Foundation in 2009.[11] Sketching out the dimensions of climate crisis, it concedes that in industrialized economies, the main emission sectors are: buildings 35 per cent, steel manufacture 27 per cent, transport 23 per cent, with cement and paper production close behind. The paradigmatic measure is that 1 ton of steel will result in 2 tons of CO_2. Meanwhile, Worldwatch cites an International Energy Agency (IEA) estimate that transitioning out of oil will cost US$45 trillion, a figure put forward by the IAE in support of the nuclear option. Worldwatch calculates that the United States and the European Union as leaders in world trade together consume approximately one third of global energy resources and emit approximately one-third of greenhouse emissions.[12] This figure contrasts sharply with estimates from the Global South claiming that its own 60 per cent of humanity produces only 1 per cent of global emissions. Worldwatch cautions against 'restarting the engine of consumption' in favour of 'fundamental green transformation'; but it also resorts to the Brundtland Commission double-speak about 'a new paradigm of sustainable economic progress'. To quote:

> properly designed carbon-markets can be effective instruments for meeting a societal goal while tapping into *the discipline and efficiencies of markets*... But markets for ecosystem protection, whether to conserve the atmosphere, waterways, or species, are not silver-bullet solutions; *the economic logic of markets may not match the scientific necessities of ecosystems.*[13]

Unfortunately, the clarity of this last sentence is not maintained throughout the *Transatlantic* blueprint for intercontinental cooperation. If 'the economic logic of markets may not match the scientific necessities of ecosystems', equally the mathematically derived logic of human engineering does not match the scientific necessities of ecosystems. This insight might have limited the heavy reliance on technological efficiency in the *Transatlantic Green New Deal*, but as things stand, the text is buoyed up with much scientific rhetoric and management hubris. Take for instance the line that 'the annual costs of reducing gas emissions to manageable levels would be around 1 percent of global GDP'.[14] While the motive behind the claim is sound, what is its empirical basis? Reliable data on aviation and agro-industrial generation of greenhouse gases is hard to get hold of; estimates of the volume of global emissions rely on informed guesswork; and the methodology of translating emissions into dollars is as arbitrary as the Gross Domestic Product (GDP) construct itself. In any event, the focus of critique here is not so much the accuracy of facts and figures, as a holistic interrogation of the ecological and social integrity of global solutions on offer.

While climate change policy goals are constantly being renegotiated, the EU has generally been ahead of the US commitments. The *Transatlantic Green New Deal* had the EU Kyoto Protocol target reducing emissions by 20 per cent from 1990 levels by 2020 or by 30 per cent if other rich nations commit to the same level. The EU is described as pioneering energy-efficient renewable technologies, directives to internalize environmental costs into prices and a cap-and-trade system supporting both labour and natural resources. By contrast, looking at global expenditure on economic stimulus packages, incentives and tax cuts, including President Obama's American Recovery and Reinvestment Act 2009, only 16 per cent would have green potential. In some respects, China, Japan, India and, particularly, South Korea, have taken the lead. Worldwatch recommends gearing up education for scientists, engineers and technicians; welfare through green jobs; a 'leapfrog' into sounder production methods; energy renewables, water harvesting, smart grids, efficient refrigerants, plug-in vehicles, fast rail and bike paths, recycled scrap and leasing in preference to purchase.[15] There is a faith in energy savings through 'dematerialization' by using nano-broadband and teleconferencing, but at the same time, Worldwatch acknowledges that computers are 'voracious users of energy' and made of toxic materials. In other words, this technology has an ecological debt, and possibly an embodied debt, largely uncounted and unspoken.[16]

Worldwatch recommends that carbon-markets and water banks be encouraged where there is no political will to fund ecosystem protection programmes directly. The Center for American Progress says that green moves could bring 2 million new jobs into the US economy. But at the same time, Worldwatch warns that competitive 'domestic-first' policies are to be avoided, and here, it seeks a coordinating role for UNEP's *Global Green New Deal* and agencies like the International Labor Organization (ILO). The *Transatlantic New Green Deal* refers to the Millennium Environmental Assessment (MEA) observation that 60 per cent of ecosystem 'services' have been destroyed since the Second World War, and its authors conclude that 'in a crowded world, whose ecosystems are already in many cases taxed beyond capacity, the continuation of conventional economic activity spells an accelerating deterioration of the natural systems that underpin environmental, human, and economic well-being'.[17] However, the classic instrumental rationality underlying the Worldwatch approach reappears in the statement that '*Ecosystems are "natural infrastructures"*'.[18]

Overall, this statement is heavily infused with androcentric denial. If the ecological conceptualization of the *Transatlantic Green New Deal* is weak, so too is its sociological lens. A new 'social contract' is on the table, but plainly the economic hegemony of the Global North dictates its terms. Thus, a number of EU states are experimenting with environmental tax revenues, yet, as the authors point out, it is important that governments do not create exemptions or, worse, subsidize bad practices:

> more can be done to rationalise current tax systems, which tend to make natural resource use too cheap and labor too expensive. Using eco-tax revenues to lighten the tax burden on labor (by funding national health or social security programs through eco-taxes rather than pay-roll taxes) would help lower indirect labor costs and boost job creation without hurting workers' interests.[19]

Nevertheless, the political perspective implied in the document is once again exclusive to the Global North. Alliances are observed in the EU between the trade unions and environmental NGOs. And indeed, these can do important work in skills training and support for newly jobless workers. In the US, the Sierra Club, the United Steelworkers Union, the National Resources Defence Council, Communications Workers and Service Employees are talking. But according to Worldwatch, the only other constituencies needing to be brought to the table are 'consumers and business'! Given that entrepreneurial interests shape the entire deal, it is perhaps no surprise to see business getting in a second time round as a 'special interest group'. This is exactly what occurred at the Rio Earth Summit in 1992 when Agenda 21 was composed.

There is no attention to structural differences in opportunity or in skill by sex/gender, race or class. The unmeasured economic input of the domestic labour sector and, in parallel vein, accumulation based on cheap resourcing of the Global South are each bracketed out. This is tantamount to leaving the voices of 80 per cent of humanity out of consideration. The only moment when Worldwatch's intercontinental brief comes close to acknowledging the existence of these meta-industrial labour groupings at the periphery of capitalism is when ethanol is rejected as an energy alternative because such crops take food-growing land away from peasant farmers. In fact, the positive climate mitigating effects of self-sufficient provisioning in the Global South seem to be fully grasped, even though it is noted that 'environmentally friendly activities . . . are often more labor-intensive than "brown" capital intensive industries'.[20] Unfortunately, this statement – compatible with a genuinely reproductive economy – is only made in passing. From the eye view of the regular professional consultant, the meta-industrial sector is simply Other, with no active economic or political agent observed at the periphery. Is it any surprise that deals built on existing social contradictions end up exacerbating existing ecological contradictions?

Australia

Yet another *Joint Statement: Towards a Green New Deal* would be issued by the Australian Conservation Foundation, the Council of Social Services, the Climate Institute, Property Council, the Australian Council of Trade

Unions, the Australian Green Infrastructure Council and the Institute of Superannuation Trustees.[21] These were all familiar personae bar the mysterious Australian Green Infrastructure Council (AGIC):

> AGIC is a company formed by a group of industry professionals from engineering, environmental, planning, legal, financial, and construction backgrounds, working in both private and public organisations related to infrastructure... Its members aim to deliver more sustainable infrastructure by driving market transformation through education, training, advocacy and a sustainable rating scheme for infrastructure projects.[22]

Prominent AGIC members included the environmental consulting firm GHD and expert tunnel builders Snowy Mountains Engineering Corporation. Australia's single-most powerful corporate lobby, the Minerals Council, was noticeably absent from among the *Joint Statement* signatories; but absent also was the Women's Electoral Lobby as well as any indigenous organizations.

Omissions of the latter political voices skew the *Joint Statement* in a particular way, its focal points being thoroughly productivist. The formula is:

1. *retrofitted buildings* to enhance energy and water efficiency carried out nationwide in residential, commercial, and public sectors; assistance for low income people as the first to undertake household efficiency audits.

2. *sustainable infrastructure* like public transport and freight rail and small renewable-energy installations – solar, wind, geothermal – to reduce the carbon footprint; special attention to the construction industry and materials sector.

3. *green industries* for the manufacture of internationally competitive new products and services, projecting 500,000 green jobs, with an 'immediate effort invested in green skills for Australia's trades men and women'.[23]

Meanwhile, as the mining lobby pressures the federal government to dilute its Carbon Emissions Reduction Scheme (CPRS), the government expands new permits for coal and gas mining.

The *Joint Statement* is understood as a 'job stimulus package' to build prosperity and insulate the Australian economy from future shocks, but if the economy itself is treated as an actor, the very real historical agency of banks and share traders is not acknowledged. And while the economy may need to be insulated from shocks, the authors do not concede that the ecosystem might also need such protection, not least because human bodies are in continuous metabolic exchange with it. Just as in the *Transatlantic Green Deal* where social justice becomes an employment ratio, so here, the ecosystem translates as 'energy efficiency'. By androcentric logic, Nature is conceptualized as a resource, reduced to a numeral and objectified as 'out there'. So too, it is claimed that energy efficiency has 'value' because it will 'reduce the $ cost' of the CPRS. This plan is described as generating 'both technology push and market pull' – which is to say that the business sector will be rewarded from both the turnover in green construction and new profits from emissions trading.

The Australian *Joint Statement* considers the simultaneous decrease of carbon pollution and increase of healthy green industries to be a 'double dividend' of 'natural and social capital'. Capitalist, indeed neoliberal reasoning and 'domestic competitiveness' also mark the Australian Conservation Foundation and the Australian Council of Trade Union assertions that 'Australia's ambition should be to capture a quarter of a trillion dollars of industry share in what will be a global industry [in green jobs] worth almost US$2.9 trillion dollars'. This is a clear commitment to export-led growth and international free trade in efficient technologies. The priorities are urban consumerism, manufacture and exchange value.[24] There is no attention to employment options in landscape restoration projects, despite the regenerative metabolic value of such work. Agriculture is passed over despite the fact that agro-industry has massive emissions – and despite the fact that sustainable small-scale farm employment and local food sovereignty would be highly desirable. At the same time, Green Party research was cautioning that even the steady-state Index of Sustainable Economic Welfare (ISEW) or measures like the Genuine Progress Indicator (GPI) rest on the incoherent GDP construct.[25] Additionally, ecofeminists such as Marilyn Waring have carefully detailed the systemic errors introduced by sex/gender illiteracy in prominent statistical indices like the United Nations System of National Accounts (UNSNA), ISEW and GPI.[26]

US Democrats

As the twenty-first century unfolds, the Green New Deal idea has ignited a second wave of proposals. One widely discussed option is advanced by US Democrat Rep. Alexandria Ocasio-Cortez in conjunction with Sen. Ed Markey. This draws inspiration from an earlier but more radical US Green Party model as well as from Bernie Sanders's visionary but unsuccessful socialist presidential campaign.[27] EcoMarxist theorist John Bellamy Foster points out that while the Green Party deal was military focused and anti-imperial, Sanders' focus was on eliminating corporate subsidies and taxes.[28] The 2019 Ocasio-Cortez GND Resolution to Congress was well inside the Keynesian capitalist welfare tradition, with heavy reliance on public investment in provisions already enjoyed by citizens of democracies like Denmark or New Zealand. Nevertheless, the agenda was sensitive to communities of colour, covering security of employment and retirement conditions; affordable health care and clean environments; free education at all levels; smart power grids and renewable energy with zero emissions; affordable homes and building retrofits; accessible public transport and high-speed rail; non-polluting industry; and sustainable agriculture. Family farms and rural jobs were supported to help develop a new food system but problems like women's unpaid labour and domestic violence got passed over. Another Ocasio-Cortez omission was preparation for refugees, as US overseas activities, including technology transfer, displace community livelihoods in neighbouring states. Moreover, political economists Anna Sturman and Natasha Heenan have observed that

> Soon after Ocasio-Cortez presented the GND in the US, the Indigenous Environment Network (IEN) pointed out that a program involving the orientation of the economy towards the care of people and planet, not to mention large-scale shifts in land use, was missing any reference to meaningful consultation with Indigenous Peoples.[29]

The Democrat plan is endorsed by 350.org, Greenpeace, the Sierra Club, Extinction Rebellion and Friends of the Earth. Curiously, the Sierra Club, the Natural Resources Defense Council, the Audubon Society and the Environmental Defense Fund oppose it, as do advocates of carbon pricing, carbon capture and storage, hydro and nuclear power – as well as most Republicans. A major weakness of the US GND is its taken-for-granted ecological modernism, as the ideology of dematerialization goes unquestioned.

EU and DiEM25

In the early years of the new millennium a number of European centres were reaching for green alternatives, but the present EU position barely recognizes this work.[30] The *EU Green Deal* presented in 2020 by Commission president Ursula von der Leyen is said 'to reconcile' with the planet. This is perhaps remarkable given that the EU created in 1951 grew out of a fossil-fuel cartel, a Coal and Steel Community, with a Common Agriculture Policy for petro-farming. The Eurocentric lifestyle is fully locked into the logic of GDP growth and dependent on the Global South with massive extractivism of food produce, deforestation and minerals. What the EU deal offers is a Farm to Fork Strategy to reduce agricultural emissions; a circular economy for textile manufacture; plastics reduction; plus access to 'sustainable' investments through a Just Transition Fund. The EU climate objective of Net Zero emissions by 2050 relies on unproven mitigation technologies like carbon capture and storage. Critics observe that it leans heavily on digitalization, shifts risks from the private to public sector and undermines citizens' rights. An analysis by the Transnational Institute reveals a slew of regulatory restrictions on digitalized trading by which the EU adopts an explicitly colonial extractivist strategy in gleaning data from Global South nations to boost its own 'value chains'. EU-WTO negotiations are described as leaning into the demands of Google, Apple, Facebook, Amazon and Microsoft.[31] The EU deal also violates the 'common but differentiated responsibilities' clause of its 2016 Paris COP Agreement and would have to spend trillions to fully accommodate climate targets. In short, the *EU Green Deal* has been described as the betrayal of a generation.[32] The proposal is dismissed by the Left as opportunist and as a technocratic capitalist greenwash of economic inequality. It may also serve to confuse the rising tide of climate populism.

The question is: Can neoliberal multilateralism be transformed into international solidarity without a culturally informed analysis? For example, are movement aspirations for global trade and monetary reform, housing justice and debt reduction too linear and 'distributional'. Ted Trainer, author of a very early call for 'a simpler way', senses too much political compromise in calls to decouple GDP from environmental impacts: 'It is a mistake and waste of energy to try to get degrowth policies implemented by governments, or to try to take the state.' By its very logic, 'capitalism cannot move in a degrowth direction'. Rather he urges, in a time of planetary crisis, the task is to 'prefigure' alternatives by commoning as the anarchists have done.[33] Stefania Barca points to a tension between the Environmental Justice movement and the trend towards ecological modernization on the Left.[34] She wonders if degrowth advocates can develop an effective alliance with the worker's movement, when so many of them accept the old 1990s postmaterialist optimism of 'green growth'.[35] That approach to 'environmentality' simply sets up a neoliberal round of the tables between market efficiency and public accountability; it inheres in the UN Sustainable Development Goals as well. When trade unions and socialists in government go with a rhetoric of Just Transitions and 'climate jobs', it locks wage-labour further into capitalism. Conversely, Barca observes that 'labour' in the generic sense is an intrinsically ecological activity, but industrialization sets up a metabolic rift between a society and its environmental conditions.[36] Labour can either destroy its ecological base while producing exchange value or work with nature as 'a force of reproduction' to generate living energies. People who know how to provide this way are a global labour majority, but their sensitivity to a life-affirming metabolic value is submerged by capital. As Barca points out: the interests of all workers, women and indigenous peoples should belong with the Environmental Justice paradigm, and she regrets that Western labour movements have not been open to a more ecofeminist understanding of work and political subjectivity.[37]

A shadow Green New Deal is evolving within the Democracy in Europe Movement (DiEM 25) inspired by the Marxist economist Yanis Varoufakis.[38] DiEM25 seeks an alliance with just transition principles, degrowth and ecofeminist radicalism. It sees the mainstream notion of 'green growth' as a false solution and argues for a multiscale paradigm to pull production away from the transnational corporate sector. Thus, international supply chains and armament sales are out. A financial transactions tax would be introduced, and the World Trade Organization broadened to include human rights. Companies providing housing, water or health care would be repurchased and managed as public services. There would be a four-day work week. An Environmental Justice Commission would oversee green public works and send jobs lost from dirty industry into recycling and habitat restoration. Its website lists the following heads: Living Costs, Natural Disasters, Job Security, Public Services, Quality of Life, Involuntary Migration, Our Future. An agrarian transition would be based on food sovereignty. Standards and payments would be specified for care work, although it remains to be seen whether traditional power relations around reproductive labour would be opened up to encourage cultural reflexivity about sex/gender.

DSA-USA

Scientific opinion from the Inter-Governmental Panel on Climate Change (IPCC) calls for 'social transformation' in order to achieve greenhouse gas reduction to 1.5 degrees Celsius, the Democratic Socialists of America (DSA) takes this further. The DSA judges its country to be in an existential crisis that exceeds ecological breakdown. 'Deepening inequality, suppressed democracy, precarious jobs, sex/gendered and racialised violence, border hostility, and endless wars make up the terrain on which climate change destabilisation will be unleashed.'[39] A prime objective of the *DSA Green New Deal* formulation is to heal old tensions between workers and environmentalists and to mobilize the working class in 'just transitioning'. In this, the DSA enjoins the Climate Justice Alliance (CJA) and the Indigenous Environmental Network (EIN). Its new deal is described as a conversation starter for growing a multifaceted movement, rather than a blueprint as such, and it welcomes the parallel initiatives of US Democrats like Rep. Ocasio-Cortez and Sen. Markey. DSA principles set out to

- Decarbonize the economy by 2030, foster renewables and agroecology
- Achieve public control of energy, water, parks and services

- Centre worker and union roles, more jobs for restoration than automation

- Decommodify everyday survival needs with a living wage, free education

- Reinvent community with cooperatives, end sex/gendered and racialized oppressions

- Demilitarize, decolonize, build international cooperation, refugee rights

- Redistribute resources, with a progressive tax on ecocidal elites.[40]

The DSA Green New Deal is quite explicit on dismantling the mega-polluting US military-industrial complex. Thus, it would

> Divert funds away from our government's bloated militarized budget, which has nothing to do with defense of people living within American borders and everything to do with maintaining imperial dominance over other nations and capitalist control of the world's resources, until global climate funding greatly exceeds military spending. Shift subsidies to extractive, exploitative, unhealthy and militarized industry to support regenerative sectors of our new economy instead.[41]

In relation to 'endless wars', there are well over 80 million refugees walking the Earth, many attempting to enter the United States. The DSA argues that reparation payments are due for economically displaced overseas communities. DSA would terminate neocolonial green washing and intellectual property restrictions on technology sharing. New social arrangements would be ecologically compatible with local terrain, forests, grasslands and wildlife. The discussion of participatory planning and neighbourhood councils acknowledges indigenous sovereignty and the need to end sex/gendered and racialized discrimination. However, the discussion of women appears to be about 'equality rather than difference'. That is, the learned skills and insights that women – indigenous and otherwise – bring to community politics are bypassed, even as words like 'care' and 'regenerative labour' are spoken. This is a common oversight in socialist thinking, a tradition that prioritizes relations of production over relations of reproduction.

Others' Deals

Tacitly, each of the GND platforms speaks to the *Androcene* debt matrix of social, livelihood, embodied, generational, species and planetary debt. Thus a further US GND platform known as *The Red Deal* has emerged under the leadership of Oceti-Sakowin peoples, opponents of the Dakota Access Pipeline at Standing Rock. Similarly, this proposal resonates with Australian First Nation struggles of Wangan and Jagalingou peoples against the Adani coal mine at Galilee Basin.[42] The *Red New Deal* commends green jobs, free housing, health care and education, and beyond Keynesian reforms it advocates change from below. As the economic excess of the twenty-first-century Global North extends to gene edits, synthetic meat and interstellar extractivism, indigenous peoples call out the massive historical debts owed to the Global South and argue that reparations are in order. *The Red Deal* echoes the Agreement adopted by some 30,000 people at the World Peoples Conference on Climate Change and the Rights of Mother Earth, hosted in 2010 by Evo Morales of Bolivia. Indigenous communities occupy one quarter of the world's land surface, but this same area is also home to two-thirds of global biodiversity. Their traditional provisioning and governance skills focus on responsible care of that land and its species; and indeed, decolonization is essential if Life-on-Earth is to survive. At the same time, Leonardo Figueroa-Helland warns against 'the appropriation of Indigeneity and Indigenous knowledges in the service of non-Indigenous and non-decolonising (sometimes hegemonic) aims'.[43]

The implosion of global finance is sometimes put down to 'human greed', but that formulation, just like the Anthropocene label, essentializes humanity, overlooking key structural differences that shape societies across the world. In parallel vein, the environmental crisis is described as a case of 'unsecured ecological credit'; but not everyone everywhere has abused this loan or mortgaged the Earth. The masters of global finance are urging that capitalism can be made sustainable. But policymakers seeking effective strategies for reconstruction need to keep social differences upfront. Under the patriarchal-colonial-capitalist imperium, structural variables like sex/gender, ethnicity and class denote sites of deprivation; but they also denote capacities and skill sets relevant to the preservation of life-support systems. To illustrate the case: if 60 per cent of global greenhouse emissions are generated by industry, 20 per cent by transport and a fair proportion by agro-industrial enterprises, then why target housewives about saving energy in the home? This is precisely what British Petroleum and other corporates tried in Australia with a One Million Women PR-designed campaign. BP's tactic was rather ironic given that women activists and theorists around the world had already outlined an eco-sufficient politics decades ago with their 'subsistence perspective'.

In a 2020 review of GND thinking for the Transform! Europe organization, Julia Marti Comas advanced a refreshing idea for regrounding policy: 'Placing the focus on popular ecofeminism already in existence . . . value proposals that set out from the "here and now"'.[44] In fact, a policy brief from a feminist Action Nexus consisting of WEDO, WWG-FFD, FEMNET and PACJA does just this with their document *A Feminist and Decolonial Global Green New Deal*.[45] The document makes a thorough structural analysis and immanent critique of the global economy offering proposals on common but differentiated responsibilities, technology transfer and intellectual property, non-discriminatory trade policy, investment in public services, multidimensional debt, financialization and tax justice. However, the women's' argument reframes the GND debate by claiming that 'feminist macro-policy puts the care-economy at its centre'. Their position builds on standard liberal feminist documents like CEDAW and the Beijing Platform for Action but goes deeper with a critique of power/knowledge and the social construction of *terra nullius*. Bhumika Muchhala describes it this way:

> The dominant neoclassical economic discipline is one out of many possible economic theories and ideas in a spectrum that is pluralistic and heterogeneous. That is why we must ask: Who is producing which 'knowledge', and what are the vested interests of these actors? Whose histories are read in textbooks and whose philosophies, theorems and methodologies are taught in schools and university curriculums?

> The colonial construction of humanity is that of a rational and objective [masculine] individual who is separate from and superior to Nature. Two historically propagated falsehoods take centre stage: Nature is claimed 'dead' and land is proclaimed 'empty'.[46]

Thus, in addition to structural and distributional reforms of the global financial system and 'the ideology that shapes nation states', this New Deal calls for holistic 'reparations' to make good the historical debts of the patriarchal-colonial-capitalist era. These qualitative measures would support smallholder agriculture and natural resource regeneration. At this point in time, the policy brief appears unresolved over whether to 'reform or delink' from the *Androcene* mega system.

13

The 2030 Agenda
Sustainable Development Goals

As First Nation peoples say: We do not inherit the Earth from our ancestors, we borrow it from our children. So the 2030 Agenda, consisting in 17 Sustainable Development Goals (SDGs) and 169 targets, was launched by the United Nations and heads of state in September 2015. It was described as a universal plan of action for 'people, planet, and prosperity' to take effect over the next fifteen years. The intention behind the SDGs is to protect the Earth from further degradation including climate change and to manage natural resources for sustainable production, in order to support the needs of present and future generations. To quote the UN Sustainable Development Knowledge Platform, *Transforming our world: the 2030 Agenda for Sustainable Development* at para 14:

> There are enormous disparities of opportunity, wealth and power. Gender inequality remains a key challenge. Unemployment, particularly youth unemployment, is a major concern. Global health threats, more frequent and intense natural disasters, spiralling conflict, violent extremism, terrorism and related humanitarian crises and forced displacement of people threaten to reverse much of the development progress made in recent decades. Natural resource depletion and adverse impacts of environmental

degradation, including desertification, drought, land degradation, freshwater scarcity and loss of biodiversity, add to and exacerbate the list of challenges which humanity faces.[1]

Fixing Poverty

The SDGs aim to end poverty by 2030. In line with the UN Charter and Universal Declaration of Human Rights, equality and the empowerment of women, children, the disabled, migrants and least developed nations is emphasized. A Global Partnership and finance for new technologies are advocated as means for enhancing the protection and care of peoples. The SDGs are designed to build on an earlier set of Millennium Development Goals (MDGs). However, these were unsuccessful.[2] The SDGs now broaden the MDGs by attempting to integrate 'the three pillars of sustainable development – economic, social, and environment', but already this statement reveals tendencies that compromise the UN programme. The policy term 'sustainable development' is widely considered an oxymoron these days. Conventional wisdom and academic disciplines falsely treat the economy as distinct from society or falsely treat the economy and society as distinct from ecology. But silo thinking is unhelpful to social change.

As ecological feminists point out, silo thinking stems from the Humanity/Nature dualism, keystone and central dogma of the patriarchal-colonial-capitalist imperium that now paves the globe. As Marxists say: capitalism is an inherently contradictory system of production. Thus the UN's famous 1980s Brundtland Report erred by promoting economic 'growth and trickle down' benefits as the solution to poverty. What was overlooked was the fact that growth or capital accumulation by some depends on appropriating the resources and labour time of Others – namely workers, indigenous peoples and women. Anthropologist Jason Hickel calculates that if poverty is measured by the $5 a day marker, then more than 60 per cent of humans are poor. Hickel adds that 'Given the existing ratio between GDP growth and the income growth of the poorest, it will take 207 years to eliminate poverty with this

[SDG] strategy, and to get there we will have to grow the global economy by 175 times its present size.'[3]

The 2030 Agenda is not just unrealistic but also undemocratic. The goals are to be realized by growing GDP, increasing market liberalization and free trade as well as according more power to the World Trade Organization (WTO). Food commodity markets are to be stabilized, but regulation of the international banking system is not tackled in the SDGs; nor is corporate tax evasion mentioned. The approach to foreign debt repayments tends to be one of restructuring investment promotion regimes for least developed countries. The UN has been increasingly open to private sector interests since the 1980s, and the World Business Council for Sustainable Development (BCSD) was especially active in drafting Agenda 21 at the 1992 Rio Earth Summit. By the time of Rio+20 in 2012, the private sector with its public relations agencies had enlisted big NGOs and ambitious academics in multi-stakeholder consultations to author a 'green economy' vision known as *The Future We Want*.[4] Meanwhile, at Davos in the Swiss Alps, world business leaders were putting forward a major Global Redesign Initiative.[5] European moves towards 'Earth System Governance' as a form of neoliberal constitutionalism have been gaining ground, even seeking formalization as a regular SDG.[6]

A Global Partnership of corporations, government and UN bureaucrats known as the High Level Political Forum on Sustainable Development is said to direct the SDGs through a body that taps into the General Assembly, Economic and Social Council (ECOSOC) and a Multi-stakeholder Forum on Science, Technology and Innovation. The neoliberal PPP or public-private partnership model is becoming a UN norm. Implementation and financing of the SDGs is voluntary, and targets are understood to be aspirational. Annual reviews of progress will be provided by the UN as a 'public service'. So Knowledge Platform Para 48 explains:

> Indicators are being developed to assist this work. Quality, accessible, timely and reliable disaggregated data will be needed to help with the measurement of progress and to ensure that no one is left behind. Such data is key to decision-making.[7]

To defer to the ecoMarxist James O'Connor here, if the 'first contradiction' of capitalism tells how social deprivation is an inevitable outcome of economic growth, the 'second contradiction' adds ecological depletion to the mix.[8] The second contradiction shows up in tensions between economy and ecology, because capital accumulation undercuts the very materiality of environmental conditions that make it possible. It is nonsensical to speak of sustainable development while advocating continued extractivism, rising GDP and expanding global free trade. In the words of degrowth theorist Jason Hickel: 'Right now global production and consumption levels are overshooting our planet's capacity by about 50 per cent each year.' He goes on,

> let's say that poor countries manage to grow incredibly fast, and quickly catch up to the average high-income-country. According to data provided by the Global Footprint Network, we would need at least 3-4 Earths to sustain this level of production and consumption – and that's assuming that the already-high-income countries slow their present rates of growth to zero, which they show no sign of doing.[9]

The need for a deep epistemological shift in how knowledge of the environment is constructed is implicit when the Marxist geographer David Harvey writes:

> the spatio-temporality required to represent energy flows through ecological systems accurately . . . may not be compatible with that of financial flows through global markets. Understanding the spatio-temporal rhythms of capital accumulation requires a quite different framework to that required to understand global climate change.[10]

High-Tech Designs

If SDG funding is voluntary and rather loosely described by the UN, *Transforming Our World: The 2030 Agenda for Sustainable Development* lays out the role of technological innovation in precise detail. Plainly, the Knowledge Platform at Goal 7a suggests an investment boom is anticipated.

By 2030, enhance international cooperation to facilitate access to clean energy research and technology, including renewable energy, energy efficiency and advanced and cleaner fossil-fuel technology, and promote investment in energy infrastructure and clean energy technology.

Developing countries, in particular, are targeted for this economic expansion and, note well, new information technologies are seen as indispensable to it. In addition, para 70 announces:

We hereby launch a Technology Facilitation Mechanism which was established by the Addis Ababa Action Agenda ... [comprising] ... a United Nations Interagency Task Team on Science, Technology and Innovation for the SDGs, a collaborative Multi-stakeholder Forum on Science, Technology and Innovation for the SDGs and an on-line platform.

The Task Team will draw on existing resources and will work with 10 representatives from the civil society, private sector, the scientific community, to prepare the meetings of the Multi-stakeholder Forum on Science, Technology and Innovation for the SDGs, as well as in the development and operationalisation of the on-line platform, including preparing proposals for the modalities for the Forum and the on-line platform. The 10 representatives will be appointed by the Secretary-General for periods of two years. The Task Team will be open to the participation of all UN agencies, funds and programmes, and ECOSOC functional commissions and it will initially be composed by the entities that currently integrate the informal working group on technology facilitation, namely: UN Department of Economic and Social Affairs, United Nations Environment Programme, UNIDO, United Nations Educational Scientific and Cultural Organization, UNCTAD, International Telecommunication Union, WIPO and the World Bank.

The question arises: Will financialization assist the implementation of SDG goals or will SDG goals assist the implementation of financialization?

The 'second contradiction' of capitalism describes how this irrational development model cuts off its very ecological base as it proceeds through cradle to grave cycle of extraction-transport-smelting-transport-manufactur e-transport-marketing-transport-operation-transport-repair-transport-wastepit. Then add the ongoing energy drawdowns in the daily use of high-tech equipment and a distinctly unsustainable toll of erosion, toxicity, water wastage and atmospheric emissions from IT. Plainly this flies in the face of other SDGs as well. However, silo thinking in academia – the hegemonic separation between economics, sociology and ecology – means that the transdisciplinary problem-solving needed to remedy these androcentric contradictions is not available. The impasse is clearly demonstrated in the current reliance on abstract market mechanisms to rectify global warming. A parallel indictment of this incoherent UN programme is offered by William Easterly, former World Bank economist, who describes the SDGs as 'Senseless, Dreamy, and Garbled . . . The SDGs are to monitor the attainment of goals that cannot be monitored or attained . . . [let alone] . . . financed by unidentified financing'. In Easterly's view the main value of the SDGs is in

> reaffirming the importance of people's right to self-determination . . . The decline and fall of the pretensions of foreign aid only tells us not to put our hopes in UN bureaucrats or Western experts. We can put our hopes instead in the poor people we support as dignified agents of their own destiny.[11]

Climate Finance

This point brings us to the nub of the argument, which is about getting to a decolonized ecological future – a socio-economy of *buen vivir* based on people's sovereign intelligence in managing their habitat, livelihood and even climate change strategy. Thus, SDG climate Goal 13 states:

> 13.1 Strengthen resilience and adaptive capacity to climate-related hazards and natural disasters in all countries
>
> 13.2 Integrate climate change measures into national policies, strategies and planning
>
> 13.3 Improve education, awareness-raising and human and institutional capacity on climate change mitigation, adaptation, impact reduction and early warning.

13.a Implement the commitment undertaken by developed-country parties to the United Nations Framework Convention on Climate Change to a goal of mobilizing jointly $100 billion annually by 2020 from all sources to address the needs of developing countries in the context of meaningful mitigation actions and transparency on implementation and fully operationalize the Green Climate Fund through its capitalization as soon as possible.

The SDG climate programme calls for a maximum average global temperature increase of no more than 1.5 degrees Celsius above pre-industrial levels, and it relies on standard capitalist measures of mitigation, adaptation, finance, tech transfer and capacity building. But this kind of environmental management means the manipulation of often *ad hoc* abstract indicators by national bureaucracies and is seen to require billions in capital outlay for technology innovation, loans and purchase. In following the Inter-Governmental Panel on Climate Change (IPCC) conceptualization of global warming with planetary-level parameters, the SDG plan shifts climate action away from Peoples' capacities to act – that is to act in any way other than as consumers.

Do world leaders addressing the IPCC Framework Convention at the regular Conference of Parties (COP) agree on the recommended carbon emissions target of 1.5 degrees Celsius? No: and the charade is ongoing. The 2015 Paris COP25 was a case in point, with activists reporting business as usual offering a polite 'we'll get back to you'. Class warfare aside, it is plain that climate politics goes nowhere as long as peoples movements remain locked into debates over arithmetic and pricing. It is time to reset the start line for climate struggles in a place that transcends silo thinking and its reductionist episteme. In Paris, serious efforts to reduce emissions were postponed to 2020, with Net Zero targets to be realized sometime after 2050. The democratic principle of common but differentiated responsibilities for rich and poor nations was replaced by undefined gestures called Intended Nationally Determined Contributions. The urgent need to control emissions from mining, animal agriculture, deforestation, aviation and shipping was sidelined. Renewables were advocated but on par with ineffectual market solutions, like offsets and carbon trading, and risky solutions, like fracking, nuclear power and geoengineering. By the time of the 2023 COP28 in Dubai, little had changed.

The profiteering commodity-based lifestyle advanced by international elites in Global North and South means that 10 per cent of the world population puts out 60 per cent of greenhouse gases. Yet a UN beholden to the corporate sector is incapable of making affluent industrial nations accept historical responsibility for this crisis. It seems fair for the Global South to demand a polluter pays principle covering liability for reparations from the North if their livelihoods and jobs have been lost through industry-generated climate impacts. But legally binding references to human and specifically indigenous rights, intergenerational equity and food sovereignty were even moved at one point from a text Agreement to a text Preamble. At the Paris meeting, environmental deregulation was promoted by default, in the enthusiastic anticipation of free trade agreements like TTIP and TPP.[12] Scholar and social justice activist Patrick Bond points out that the indifference of major powers was signalled in a 2015 announcement of no decarbonization before 2100. Bond continues to make a powerful immanent critique of the neoliberal market responses to climate favoured by the United States, the EU and subimperial BRICS coalition – Brazil, Russia, India, China, South Africa – now joined by several other once-colonized states, as well as Saudi Arabia.[13]

False Consensus

To summarize: the international model of green capitalism carried forward in the declaration, *Transforming Our World: The 2030 Agenda for Sustainable Development*, reveals the following flaws in the Sustainable Development Goals (SDGs).

- No analysis of how the structural roots of poverty, unsustainability, and multidimensional violence are historically grounded in patriarchal-colonial-capitalist institutions like state power, corporate monopolies, neocolonialism

- Inadequate focus on direct democratic governance with accountable decision-making by citizens and self-aware communities in face-to-face settings

- Continued emphasis on economic growth as the driver of development, contradicting biophysical limits, with arbitrary adoption of GDP as the indicator of progress

- Continued reliance on economic globalisation as the key economic strategy, undermining people's attempts at self-reliance and decolonial autonomy

- Continued subservience to private capital, and unwillingness to democratise the market through worker–producer and community control

- Modern science and technology held up as social panaceas, ignoring their limits and impacts, and marginalising Other knowledge

- Culture, ethics, and spirituality sidelined and made subservient to economic forces

- Unregulated consumerism without strategies to reverse the global North's disproportionate contamination of the globe through waste, toxicity, and climate emissions

- Neoliberal architectures of global governance becoming increasingly reliant on technocratic managerial values by state and multilateral bureaucracies.[14]

This 2030 framework of SDGs, now global in its reach, is a false consensus. For instance, its call for 'sustained economic growth' contradicts the majority of other SDGs. Once 'development' is understood as environmentally toxic, then 'sustainable development' is simply an oxymoron. Periodically, public debate revives old-style 'common-sense' talk about overpopulation as the key.[15] By this ruse, the Global North's responsibility for pollution levels is deflected on to the Global South. On the other hand, if 60 per cent of humanity in the two-thirds world is responsible for only 1 per cent of global warming, why talk about population?

Water Is Life

Finally, to revisit the New Water Paradigm in the context of SDGs. Here the compelling analysis of hydrologist Michal Kravcik is relayed by two young activists from the Tamera Community in Portugal:

> the annual loss of 50,000 square miles of forest and the additional soil sealing of 20,000 square miles per year have reduced the water that is able to circulate in small rainwater cycles . . . [The modelling] estimates that, throughout the last century, around 8900 cubic miles of water for these climatically crucial cycles was lost. This equals three times the water volume of Lake Superior. If you calculate the effect this has on the oceans, you end up with a sea level rise of around four inches . . . Rainwater and humidity are vital parts in the cooling system of the atmosphere. During evaporation, a gallon of water spends 2.5-kilowatt hours of solar energy. The loss of significant amounts of water and the desiccation of soil and of air therefore produce potential heat, which amounts to, as Kravcik calculated, the gigantic figure of 25 million-terawatt hours. This is 1600 times more heat produced annually than all of the planets' powerhouses combined.[16]

The secret of this integrative paradigm is working closely with biodiversity and soils to rehydrate land and subterranean aquifers. The model is inexpensive, with hands-on water restoration technologies using local stone, wood and plants, designed and carried out by neighbourhoods and communities. Its methodology is synergistic: that is to say, it simultaneously restores livelihood, provides jobs and education; it grows solidarity, cultural autonomy, empowerment and spiritual renewal. This analysis not only provides an integrative reading of climate change but also negates the second contradiction of capitalism by reclaiming climate as an ecological not an economic problem. The new paradigm indicates that climate solutions through repair of the small water cycle are readily available to people whether they live in rural or urban spaces. Versions of this paradigm are corroborated in Australia, China, India,

Canada, the United States and Europe.[17] Canadian water policy specialist Maude Barlow commends the model, and it validates Via Campesina's decolonial claim that their small-scale provisioning is 'cooling down the Earth'.[18]

Rather than be sidetracked by neoliberal silo thinking and carbon pricing, climate justice action in the Global North might simultaneously revive environmental politics and support decolonial struggles by adopting a holistic water-soil-biodiversity strategy. Bringing climate politics down to Earth by celebrating people's sovereign intelligence joins each of the SDGs (2015) together, for access to water is fundamental to the realization of almost all of them.

> Goal 1 – End poverty in all its forms everywhere.[19]

Comment: Current models of development destabilize both local and global water cycles. Poverty cannot be ended without adequate access to water; fresh water is a human right; people suffering water scarcity number over a billion.

> Goal 2 – End hunger, achieve food security and improved nutrition and promote sustainable agriculture.

Comment: Growing seasons become irregular if the water cycle is disturbed. Industrial production moves people off healthy farmlands and leaches toxic chemicals into streams and soils. Soils rely on fresh water to keep organic content active and fertile; food cannot be cultivated or livestock cannot be raised without water.

> Goal 3 – Ensure healthy lives and promote well-being for all at all ages.

Comment: Clean drinking water is essential for sustainable livelihoods; for health and sanitation; 80 per cent of sickness around the world is due to contaminated water.

> Goal 4 – Ensure inclusive and equitable quality education and promote life-long learning opportunities for all.

Comment: When parents and children are forced to spend time securing basic needs like carrying water to their homes from afar, time for educational opportunities is lost.

> Goal 5 – Achieve gender equality and empower all women and girls.

Comment: Adequate water and sanitation facilities are basic to women's reproductive health and physical safety.

> Goal 11 – Make cities and human settlements inclusive, safe, resilient and sustainable.

Comment: The new water paradigm ensures that water is not wasted at home or through industrial production. It avoids heavily engineered drainage channels which cause urban floods, pollute oceans and deplete the water table.

Goals 14 and 15 already recognize that climate is a complex non-linear system closely implicated with the functioning of water bodies; and that protection of vegetated ecosystems will help prevent desertification and biodiversity loss. But do they indicate that resilience means healing the Humanity/Nature divide and its dualist hegemony? Predictably, the ecological agency of water is backgrounded under the *Androcene* development model. Hence the need for goal 6.3.

> Goal 6.3. – By 2030, improve water quality by reducing pollution, eliminating dumping and minimizing release of hazardous chemicals and materials, halving the proportion of untreated wastewater and substantially increasing recycling and safe reuse globally.

The UN call for cross-boundary management and innovative technologies for remediation exacerbates both the first and second contradictions of capitalism.

> 6.a. – By 2030, expand international cooperation and capacity-building support to developing countries in water- and sanitation-related activities and programmes, including water harvesting, desalination, water efficiency, wastewater treatment, recycling and reuse technologies.

At first glance, such schemes may sound sensible, but it is essential to understand the top-down political economy of water in today's world. As Maude Barlow writes:

There are ten major corporate players now delivering fresh water services for profit. Between them, the three biggest – Suez and Vivendi [now renamed Veolia Environment] of France and RWE-AG of Germany – deliver water and wastewater services to almost 300 million customers in over 100 countries . . . The World Bank serves the interests of water companies both through its regular loan programs to governments, which often come with conditions that explicitly require the privatization of water provision . . . [Yet] . . . private water companies have jacked-up water rates and cut-off services to those who can't pay the bills, while reducing water quality and refusing to make investments for the improvement of infrastructure such as leaky pipes.[20]

Not long back, a Suez Environmental research team manager was sponsor and keynote for an Earth System Governance conference on SDGs at the University of Rheims.[21] Should this be understood as corporate social responsibility? Or is the writing already on the wall? On a more optimistic note, let's not forget that it was the enforced privatization of water that opened the way to a people's uprising in Bolivia and progressive climate alliances between workers, indigenous peoples and women. The post-development politics of *buen vivir* is inspirational; but here and now, on the ground, there is a long way to go. In the north of Peru, communities strike against yet another venture from the Global North – the AntaKori mine owned by Regulus Resources, Canada. They can hardly believe that 'In the dead of night, during the pandemic, the miner barged into the territory'. But water springs from this Chancay Alto Basin supply an entire coastal area including cities. Locals list existing mine impacts as responsible for the death of some 17,000 fish, for pastures drying up, for cattle losing hair, for sheep not producing milk. 'We used to sell milk, but now people don't want to buy it, claiming it is contaminated with heavy metals.' Against state indifference, even militarization, people refuse the imposts of extractivism on their territories and bodies, and their political agenda is very clear. 'We need a grassroots, parity-based, plurinational, and sovereign constituent assembly process.'[22]

14

The Smart ReSet
A Biopolitical Turn

Modern history has been driven by a series of industrial inventions – the eighteenth-century steam engine, nineteenth-century electricity and twentieth-century computing. Now, twenty-first-century digital applications include artificial intelligence (AI) and robotics, the Internet of Things (IoT), and soon enough, an Internet of Bodies (IoB). This Fourth Industrial Revolution (4IR) has reanimated the patriarchal-colonial-capitalist system. In the global South, land dispossession, especially for mining, intensifies; in the global North, capital accumulation is designed around digiitalized services, rent-seeking and financial austerity. If states like Venezuela, Ecuador, and Bolivia once flourished as the voice of peoples' movements; subject to new outside pressures, they have now fallen into disarray.[1] Meanwhile, the Eurocentric development model of agro-industry, dam construction and transcontinental supply chains continues to shake planetary ecologies. Could this environmental breakdown have led to the viral mobility that caused the worldwide Covid-19 outbreak in 2020, as Rob Wallace and others, including Walden Bello, have argued?[2] Others speculate that the virus was an accidental by-product of genetic engineering experimentation for medicinal purposes, or even a particulate escape from a biological weapons lab. The original source of Covid remains to be clarified.[3]

Nevertheless, there are sociological ties between the ecomodernist move to digitalization of everyday life and the pandemic phenomenon. One is the convergence of genetic engineering technology and information technology – a development monitored by the international NGO known as Ecology, Technology, Convergence (ETC) since the 1990s. A second link is specific, in that during the pandemic people were urged to purchase new digital apps to monitor infection hot spots with their cell phones.[4] In fact, at this time, biopolitical surveillance was leveraged with everything from card swipes at expressway tolls to customer email lists in coffee shops. Meanwhile, openings for new vaccine research were widely pursued by public-private partnerships looking for new intellectual-property, so a new awareness of genetic engineering imploded. Yet many people were fearful of compulsory injection with hastily marketed products. In regular conversations about the interplay of Big Data and Big Pharma, billionaire Bill Gates' position was questioned as both an IT entrepreneur and long-time investor in World Health Organization (WHO) promoted vaccination programmes.[5] In Africa, the Covid vaccine price was a key concern, since medicines patented internationally under trade-related intellectual property rules have proven prohibitive to communities there in the past. Meanwhile, with people redefined by state governments as 'biohazards', the pandemic, introduced a new era of digital policing. In Australia, however, androcentric logic held its own. Thus, while street rallies for Black Lives or Refugee Rights were placed under quarantine bans, football crowds were assessed an acceptable risk!

The Fourth IR

The arrival of Covid-19 allowed Big Data and Big Pharma, each a lucrative offshoot of Big Oil, to integrate their international operations. As mandatory isolation regimes suddenly became an essential feature of public life, IT communications replaced face-to-face sociability. Education was transformed into remote learning, although many teachers reacted against what they saw as corporate opportunism.

> [S]ince school closures, there has been a huge upsurge in profit making for education technology (ed-tech) companies. Commercial companies have increased their involvement in public education through ... multi-stakeholder coalitions, international organisations such as UNESCO, the OECD, the World Bank, national governments and others.[6]

Likewise, as Deane Neubauer writes in a recent Routledge anthology, the Fourth Industrial Revolution has pioneered a digitalized shift in public health care known as Health 4.0.[7] The rationale for this was explained already in 2016 by Klaus Schwab, leader of the World Economic Forum.

> We have yet to grasp fully the speed and breadth of this new revolution. Consider the unlimited possibilities of having billions of people connected by mobile devices, giving rise to unprecedented processing power, storage capabilities and knowledge access. Or think about the staggering confluence of emerging technology breakthroughs, covering wide-ranging fields such as artificial intelligence (AI), robotics, the internet of things (IoT), autonomous vehicles, 3D printing, nanotechnology, materials science, energy storage and quantum computing, to name a few. Many of these innovations are in their infancy, but they are already reaching an inflection point in their development as they build on and amplify each other in a fusion of technology across the physical, digital, and biological worlds.[8]

The 'fusion' of digital and biological is a disturbing move into biopolitics, given the violence on living processes that the Fourth Industrial Revolution implies. Neubauer comments that through digitalization, Health 4.0 rationalizes and operationalizes care systems to optimize efficiency and lower costs. It relies on agencies such as WHO's Global Outbreak Alert and Response System (GOARN), as people are encouraged to use cell phones or wearable monitoring devices and implants to gather data on their health status. Potentially the entire global population will then become a reference group for purposes of research and management.[9] One such technology is known as MADIP or Multi-Agency-Data-Integration-Program; another is called BLADE, standing for Business Longitudinal Data Sets. Notwithstanding this electronic screen, the Smart City will be more vulnerable to hacking attacks than any citadel in history. Questions of privacy and consent are put aside before 'public good'. Much of the collected data will be held privately, but since it is 'a product' for sale, the sources of surveillance are negotiating to keep governments at arm's length. Yanis Varoufakis, an economist with the Left's Progressive International network, describes such developments as techno-feudalism.[10] Critics from the Right, or simply apolitical folk, take to the streets with an instinctive sense of being bullied. Indeed, what is occurring here is a new enclosure movement, wherein the human body as a 'biological asset' is invaded by capital for extraction of its 'behavioural surplus'.[11] In 2024, the WHO is in negotiation for a Treaty with nation states, in order to institutionalize the practice of lockdown, whenever mandated by the WHO. Given that the WHO is a privately funded body, supported by philanthropists with interests in Big Pharma, this move furthers the corporate capture of the modern state.[12]

As digital technologies become ever more sophisticated, the commercial benefits of planned obsolescence are set in motion. The expansion of robotics, marketing algorithms, drone warfare, cryptocurrencies, nanotech and genetic engineering, fortifies commodification in an industrialized global culture based entirely on false needs. Everything from instant translation to 'virtual-try-before-you-buy' cosmetic apps is promoted as ushering in 'a higher quality of life'. The business elite, Dow, IBM and Siemens among them, along with opportunistic politicians meeting annually in Davos for the WEF, are big supporters of Klaus Schwab's Global Redesign Initiative, as are their 'sub-imperial' partners the BRICS – Brazil, Russia, India, China and South Africa. To quote Chairman Schwab again:

> In the future, technological innovation will also lead to a supply-side miracle, with long-term gains in efficiency and productivity. Transportation and communication costs will drop, logistics and global supply chains will become more effective, and the cost of trade will diminish, all of which will open new markets and drive economic growth.[13]

On the underside of this 1/0 expanse, the Fourth Industrial Revolution will bring automation, worker deskilling and poverty for the global majority and already signs of personal disempowerment and trust deficits can be read in the uneasy mood of street protesters. For investigative journalist Naomi Klein:

This is a future that claims to be run on 'artificial intelligence' but is actually held together by tens of millions of anonymous workers tucked away in warehouses, data centers, content moderation mills, electronic sweatshops, lithium mines, industrial farms, meat-processing plants, and prisons, where they are left unprotected from disease and hyper-exploitation.[14]

For profit-makers, the use of nanotechnology to re-fashion biological derivatives is seen as a boon to construction, health care and even agriculture; thus, while medical wearables monitor the user's physical condition, 'sensors' will tell the farmer when to fertilize the crop. However, according to ETC research into Sustainable Development, Technology Assessment, and Corporate Monopolies, the use of genetic redesign to make organisms behave like microbial factories will simply liquidate ecosystems. An article in the *Bulletin of Atomic Scientists* examines this same issue in relation to the pandemic.[15]

There was a moment at the start of the health emergency when it looked as though governments might wake up to the fact that the global economy is upside down. Did eco-socialists with their traditional productivist focus fare better in this respect? Not really. As ecofeminists argue 'the productive sector' weighs too heavily on 'the reproductive sector', sphere of human needs and ecological flourishing.[16] The global pandemic brought this historical truth to the surface by revealing how indispensable the regenerative labour force is. Moreover, it has long been part of women's given role to serve as society's shock absorbers.[17] It is an unspoken fact that the non-valued meta-industrial work of mothers and nurses supports the embodied thermodynamic infrastructure of capital, thereby helping maintain its economic surplus. But this political insight has not yet matured.

Internet of Things

The First, Second, Third and now Fourth Industrial Revolution reinforce the old *Androcene* psychology with its dissociated Humanity/Nature divide. But materialized electronically, the modern imperium and its value chains are more formidable than ever. The digital future and its panoptic 'platform economy' – response to an already faltering overproduction crisis – consist in a hierarchy of deals between the state and capital. In this revolving door, the rationale is that multinational companies will relocate to world cities with the most competitive IT connectivity. The nation state will gain billions from Telco businesses by auctioning wireless spectrum licences, as well as by infinite surveillance capacity over local citizens and foreign competitors alike. In return, business gains state support for tech promotion, as well as assurance of 'neutralized' regulators. Behind giant signal carriers – in the United States, AT&T; in Australia, Telstra – stand the internet conglomerates – Google, Apple, Microsoft, Amazon and Facebook. Already valued in the US$ trillions, these companies profit from the sale of gadgets and applications like the smart phone pandemic monitor and from the sale of user data. The latter innovation was a somewhat accidental discovery, as Google realized it might usefully 're-purpose the waste' sitting on its servers. Today, algebraic algorithms embedded in digital products monitor and profile the preferences, emotional concerns and sensitive medical information of private citizens. This data is mined like a 'natural resource' to on-sell to the entrepreneurial makers of new products. Often, fake online advertisements will be run to measure potential buyer interest among browsers.

The next digitalization phase, the Smart City and worldwide Internet of Things (IoT), would extend this adaptation of Silicon Valley's military-industrial technologies into the household itself. Such a development is neither constitutional, nor legislated, nor even debated, but the widespread assumption is that eventually every home will be equipped for global communication through WiFi. Access to IT is sometimes presented as a 'human right', but the IoT automatically connecting and monitoring domestic appliances like TVs and refrigerators puts people's living quarters under the electronic eye – not to mention floating street drone cameras – panopticon *par excellence*. At the same time, the layered intricacy of such telecommunications buffers ruling-class investors from accountability. The classic political ideal of 'just rule' through representative democracy is already compromised by the potential to hack wired-up electoral systems. Meanwhile, social media outlets like YouTube serve as a meme factory for PR agencies marketing persuasive commercial or political messages. Fake but titillating news spreads rapidly across continents through popular media platforms, leaving publics confused and often angry and divided against each other.

Social ordering under the *1/0 imaginary* affirms the historical entitlement of the productive sector to subsume the reproductive sector as 'a natural renewable'. Regular patterns of abuse of Others like women or racialized minorities are streamed as news or semi-pornographic infotainment. In the transhuman culture of cyber-anonymity, the Humanity/Nature abstraction hardens, even as scientists ponder the Anthropocene. In fact, for many, the latter provides a scientific rationale and call for more 'terraforming'.

A generation or two back, sociologist Wright Mills predicted the rise of a US power elite and 'permanent war economy'. In 2020, the state consolidates its powers through Twitter and Facebook, while intelligence professionals like the late Henry Kissinger and a number of former generals circulate between the public and corporate sectors, selling strategic thinking for big consultancy fees.[18] Links between privately owned IT firms and the US Department of Defense (DOD) grew considerably after the 9/11 crisis. A central figure in that move has been Eric Schmidt, formerly of Amazon but now worth US$ billions in Google shares alone. Schmidt advises the DOD as chair of its Innovation Board and as chair of the National Security Commission on Artificial Intelligence (NSCAI). Should the Smart City be described as social engineering or military engineering? In any event, given the frequency of international hack attacks, the difference between peace and war becomes blurred in a milieu of 'cyber-enabled state-based actors'. Already in 2020, the Australian Cyber Security Centre (ACSC) estimated the frequency of criminal data breaches on private citizens as one every ten minutes.[19] Harvard legal scholar Shoshana Zuboff traces the source of this mindset to US 'exceptionalism'. Noting that Congress is conflicted about regulating Big Data, she argues that democracy 'disfigures itself' as most Americans remain oblivious to the consequences of biometric control. Zuboff adds that Google Earth's App, Google City, was designed to coordinate all dimensions of urban life from telecommunications, to power supply and to traffic. Apparently, the popular game Pokemon Go was a trial run for the design and control of urban population dynamics.[20] Meanwhile, the World Economic Forum's Global Redesign Initiative favours a United Nations-backed but privately funded digital governance body. However, it is encouraging that people with past experience in multi-stakeholder forums have sent a worldwide petition to the UN Secretary-General opposing this development, and it is signed by hundreds of grassroots organizations.[21] While think tanks like the Transnational Institute are examining the social consequences of the Fourth Industrial Revolution, its biophysical impacts rarely receive attention. As noted, the Left in general has been very slow in responding to the technology question – an inherent Prometheanism perhaps?

The WEF Great ReSet clearly threatens human rights, but there are also materially embodied effects of living in a Smart City. It is not often realized that the convenience of wireless connectivity comes at a biological cost, as cities are blanketed under electromagnetic radio frequencies (EMR), largely microwave radiation. The Oceania Radiofrequency Scientific Advisory Association has a database housing hundreds of peer-reviewed studies indicating that microwave technology effects may include brain tumours, immune disorders, DNA damage, male infertility and electro hypersensitivity, among other conditions.[22] A phone is allowed to emit around 2 watts per kilogram into body tissue. Indeed, many mobile phone manufacturers provide specific recommendations in their user manuals. Apple, for example, advises maintaining a distance of at least 5 mm between the iPhone and the body to ensure compliance with RF exposure guidelines. The accepted heating standard is based on this gap between body and phone, but the reality is that phones tend to be held directly against the head with no air gap. Some forty-eight mobile phones are known to fail the current heating standard and should be withdrawn from the market.[23]

According to the US Federal Communications Commission (FCC), tower beam emissions may go up to 20 watts per channel. But with smartphone technology, fast-pulsed millimetre waves only travel short distances. So, in order to generate a long and steerable connective beam, public distribution antennas on high city buildings are arranged in synchronized arrays. Likewise, in the phone handset, the receiving antenna will use a powerful phased array. In urban settings, radiofrequency beams will be directed from Telco relay towers to nodes on local street poles about 300 metres apart. However, since the electromagnetic pulse is not strong enough to penetrate the brick walls of people's homes, repeaters may be placed on glass windows or the roof to enable signals to beam throughout the building. These millimetre wave frequencies (mmWaves) can penetrate the surface layer of skin or the sclera of the eyes, but with high data rates, modelling has shown that penetration will be much deeper. Unfortunately, the currently accepted industry heating standard treats skin as an inert substrate with no biological function, yet the skin is the largest organ of the body and it interfaces with the immune system.

Captured Agencies

Government agencies like the Australian Communication Media Authority (ACMA) have a major conflict of interest in both selling the spectrum and selecting the public health standard. ACMA adopts its guidelines from the Australian Radiation Protection and Nuclear Safety Agency (ARPANSA). However, this is a weak protective option. The standard set by engineers and physicists relies on power density and assumes that some short-term body heating is the only health concern. It is disturbing to find governments ignoring the accumulation of scientific evidence on the collateral damage of these products. From the outset, cell phones were marketed without testing for medical and environmental effects. An investigation by the New York-based journal *Nation* discovered that early research on electromagnetic radiation effects was sidelined when presented to AT&T, Apple and Motorola for comment. This was the case even as one study advised that 'The risk of rare neuro-epithelial tumors on the outside of the brain was more than doubled . . . in cell phone users . . . the ability of radiation from a phone's antenna to cause functional genetic damage [was] definitely positive'.[24] Industry rebuttals have invoked various government authorities including the US Food and Drug Administration and National Institutes of Health. In 2018 the US National Toxicology Program affirmed that microwave radiation is a possible carcinogen (category 2B), and the World Health Organization (WHO) should raise the classification to Category 2A a probable carcinogen, or Category 1, a carcinogenic agent. However, the US Federal Communications Commission is widely acknowledged as a captured agency, firmly under the control of Big Wireless – Motorola, Ericsson, Nokia, Samsung, Sony, GSMA and Deutsche Telekom.

Unsurprisingly, Section 704 of the 1996 US Telecommunications Act was devised to ensure that no environmental or public health concern would interfere with the installation of cellular antennas. Other countries have also followed this move. In 2013, President Obama asked Tom Wheeler, a leader in the trade, to chair the Federal Communications Commission (FCC) proactively, but the latter preferred to rely on private sector opinion. Telco influence now penetrates the US Centers for Disease Control and Prevention (CDC) and the American Cancer Society. More recently, Congress passed a law known as the 'Secure 5G and Beyond Act of 2020':

> To require the President [within 180 days] to develop a strategy to ensure the security of next-generation mobile telecommunications systems and infrastructure in the United States and to assist allies and strategic partners in maximizing the security of next-generation mobile telecommunications systems, infrastructure and software, and for other purposes.[25]

The states of California and Maryland have attempted to sue the US Federal government to regain regulatory control over their local environment. The state of New Hampshire has held a comprehensive commission of inquiry into microwave radiation from 5G. This inspired the Green Party of California to issue a precautionary policy statement pointing to the FCC's long-outdated regulatory standards. The state of Oregon is researching microwave technology impacts, and the Hawaii County Council has banned the introduction of 5G until it is proven safe. In August 2021, summing up a legal action by the *US Environmental Health Trust v. the Federal Communications Commission*, the US Federal Court of Appeal No 20-1025 ruled in favour of public safety concerns over radio frequency exposure limits.

Elsewhere, the governments of Great Britain, France and Israel have warned against cell-phone use by children because their neural development is incomplete. In Britain, there is talk of an Online Harms Bill based on 'duty of care'. Several European capitals, including Geneva and Brussels, and 600 municipalities in Italy, have placed a moratorium on 5G installation. Dr Jacqueline McGlade, executive director of the European Environment Agency, has asked for tighter controls, but there are differences in high places. A 2020 press release from the Council of the European Union quotes Svenja Schulze,

Germany's Federal Minister for the Environment, Nature Conservation and Nuclear Safety, saying: 'Digitalisation is an excellent lever to accelerate the transition towards a climate-neutral, circular, and more resilient economy. At the same time, we must put the appropriate policy framework in place to avoid adverse effects of digitalisation on the environment.'[26] The statement recommends 'regulatory or non-regulatory measures' to reduce the footprint of communication networks, data centres and waste disposal. It seeks to respond to both the New Industrial Strategy for Europe and the European Green Deal with its Circular Economy.

A major obstruction to progressing these matters is the Europe-based informal industry group, The International Commission on Non-Ionising Radiation Protection (ICNIRP).[27] The WHO and most national policymakers follow ICNIRP safety standards without questioning its conflict of interest. ICNIRP guidelines are weak and look only at the heating effect of microwave radiation. Russia has set guidelines one-hundred times below the ICNIRP guideline. India has set its safety limit one-tenth below the ICNIRP guideline; Israel and Poland are also treading with care. However, the Australian Radiation Protection and Nuclear Safety Agency (ARPANSA) has close ties with ICNIRP personnel, and research funds circulate between both organizations and the National Health and Medical Research Council (NHMRC). An international petition to the United Nations, WHO and European Union lists multiple peer-reviewed scientific research papers evidencing harm from microwave radiofrequencies well beyond thermal effects.[28] However, by fostering uncertainty over what is effectively a public health experiment in cell-phone technology, business moves steadily ahead with the Smart City, the household Internet of Things and soon the Internet of Bodies (IoB) with the invention of 'wearables' and surgically implanted microchips to replace keys and credit cards.[29] The ICNIRP guidelines have blurred the distinction between medical applications and consumer applications. Concerning the latter, many are quite frivolous, such as a digitalized fork that indicates if you are eating too fast; a pet food dispenser that allows you to talk to your dog remotely; and a wireless activated cradle to rock newborns. For Yuval Harari, posthumanist science adviser to the Davos WEF, the algorithm of *Homo sapiens* is definitely obsolete.[30]

New Climate Impacts

The complacency of official circles contrasts sharply with scientific assessments. For example, the University of Melbourne-based Centre for Energy Efficient Communications (CEET) estimated that

> by 2015, wireless cloud will consume up to 43 TWh, compared to only 9.2 TWh in 2012, an increase of 460%. This is an increase in carbon footprint from 6 megatonnes of CO_2 in 2012 to up to 30 megatonnes of CO_2 in 2015, the equivalent to adding 4.9 million cars to the roads.[31]

The Melbourne group strongly recommends staying with landline Telco connections. Meanwhile, researchers from Yale, MIT and Purdue University examined the ecological consequences of the rush to Zoom communications following the Covid-19 travel restrictions. Their calculations found the new technology brought an exponential growth not only in carbon emissions but also in water usage.[32] Jean-Marc Janovici, a member of the French High Council looking into planetary thermodynamics, reports that 'the carbon footprint of the global digital system is already 4% of the global greenhouse gas emissions, and its energy consumption rises by 9% per year'.[33] A 2020 article by Mathew Barton in *The Ecologist* argues that digitalization is already responsible 20 per cent of global usage emissions – a footprint comparable with the worldwide aviation industry.[34]

A future estimate is that 5G will require three times the electric power consumption of earlier digital technologies.[35] Energy is needed all the way, from running domestic computers to air-conditioned cooling of Cloud data storage centres. There are already eight million Cloud centres around the world, so as microwave technologies are rolled out for the promised Internet of Things, there will be an exponential increase in global warming impacts.[36] Use of energy guzzling crypto currencies internationally will add to this, and electric cars, which merely shift the pollution source from one to another. Many IT firms, even those collaborating with the US Department of Defense's Advanced Research Projects Agency (DARPA), support the production of solar and wind renewables. But again, given the linearity of instrumental reason, the calculation of dematerialization invariably elides the full cycle of production costs, including the sixfold pattern of debt owed to *Androcene* others.³⁷ The proposed use of geo-engineering to remedy climatic effects is another commercial venture finding its way into government policy. As the peoples' think tank ETC argues: the release of stratospheric aerosols and cloud whitening,

> due to their unprecedented scale, are both beyond the parameters of real-world scientific testing and beyond the scope of current international law . . . OECD countries and corporations are pushing for massive 'technofix' experiments rather than reducing emissions at home.³⁸

A further environmental cost of Great Reset infrastructure will result from the felling and pruning of trees to allow unobstructed 5G supply to households in urban areas. This issue has already led to street demonstrations in France.

Looking at the politics of reproduction, *Androcene* 'development' has multiple corollaries. Computer casings will continue to be made of plastic which ensures the ascendancy of Big Oil. Or, if these were to be replaced with a biodegradable fibre product, that would take land from people's food crops or climate-calming forests. Meanwhile, the mining of rare earth metals like lithium, cobalt, neodymium, silicon and coltan for digital wiring and batteries, as well as wind turbines and solar panels, is already destroying livelihoods in the Congo, China, Argentina, Australia and the United States. A medical test of Congo miners shows heavy metal loads in their urine: antimony, beryllium, fluorine, germanium, niobium, platinoids, rhenium, tantalum, tungsten and vanadium. Meanwhile, seabed mining destroys ocean species' habitat, and plans are afoot to mine rare metals on the moon.³⁹ At this point, Australian eco-activist Michelle Maloney asks: We have a Tribunal on the Rights of Nature, but is it time to consider the Rights of the Moon? Lithium batteries for hybrid and electric cars or buses are effective in storing energy and can be recharged, albeit from the nearest power station linkage.⁴⁰ But the 'greening' of transport comes with ecosystem damage like mountaintop removal, aquifer depletion and farmlands spoilt by toxic residues. The progressive spin on electric vehicles – metal and plastic monsters as they are – is another cost displacement from one source to another. The manufacture of IT components in Mexico or China further pollutes local air and water supplies, as well as damaging the eyes and reproductive organs of the mostly women assembly line workers who make them. Global supply chains for component delivery by cargo ship, air, rail and road transport demonstrate a massive energy drawdown. At the other end of the product cycle is the disposal of non-biodegradable electronic and plastic devices. Jon Marshall and colleagues cite a figure of 142,000 computers and over 416,000 other mobile items discarded daily by consumers in the United States. An estimate of European e-waste is given as twelve million tonnes annually.[41] By the logic of colonization, this e-waste gets to be exported by the global North to the global South. If village peoples' livelihood lands, say in Ghana or India, have been appropriated for 'development' by agro-industry, their communities may have no choice but to make a living as waste pickers. But dumped e-waste results in lead, cadmium, mercury, arsenic, beryllium or brominated flame-retardants leaching from landfill into soil and water tables. Those with no other means of subsistence, usually women and children, attempt to recycle such chemically poisonous garbage for cash. This occurs despite a Basel Convention on the Control of Transboundary Movements of Hazardous Wastes and their Disposal.

What would Earth System Governance enthusiasts suggest as an ecomodernist solution here?[42] Who has a right to set the global thermostat? The ETC group observes that geo-engineering appears to violate the Convention on Biological Diversity, the UN Environmental Modification Treaty (ENMOD); the International Covenant on Economic, Social and Cultural Rights; and the International Declaration on the Rights of Indigenous Peoples. In the words of Tom Goldtooth from the US Indigenous Environmental Network (IEN), the 'technological nightmare has been forced on us by colonial regimes for five centuries'. Given that less than half of the world population uses IT, this ecological and livelihood debt should jolt reformist Green New Deal platforms that currently bypass the material imposts of urban industrialization. In the language of ecoMarxists, capitalism sets up a metabolic rift as it severs ecosystemic flows between natural landscapes and developed towns.[43]

In 2020, the entire planetary surface is being urbanized under microwave communications. 'A recent IDC [International Data Corporation] study claims that by 2025, worldwide data traffic will have grown by 61 per cent to 175 zettabytes, with roughly 75 percent of the population having at least one data interaction every 18 seconds.'[44] While modern city dwellers walk and sleep in an invisible cloud of electromagnetic radiation generated by antennas, routers and cell phones, the unintended consequences of this cybernetic environment are still being discovered. For example, in a newsletter from Japan's Harmonix Corporation, company founder Shigeaki Hakusui writes about the problem of oxygen capture. His firm is a specialist manufacturer of the GigaLink 60GHz digital radio system used for high-speed wireless communication. Hakusui notes how weather conditions affect electronic transmission, but he gives special attention to how 5G radiofrequency (RF) communications absorb atmospheric oxygen. To quote:

> Due to the high levels of atmospheric RF energy absorption, the millimetre wave region of the RF spectrum is not usable in long haul, wireless communications segments.... While oxygen absorption at 60 GHz severely limits range, it also eliminates interference between same frequency terminals.... Theoretically, 100,000 systems operating at 60 GHz can be co-located in a ten square kilometer area without interference problems.[45]

Does Hakusui imply by 'oxygen capture' that CO_2 or something else is produced in its place? This is not made clear. Nevertheless, Hakusui notes that the US Federal Communications Commission (FCC) has allocated the millimetre wave RF spectrum from 57.05 to 64 GHz for unlicensed use. The question that begs to be asked is with a '100,000 systems operating across a ten square kilometre area', how much oxygen is removed daily and what effect does this have on humans living under these systems? Why are global leaders not demanding research into this risky science?

Colonizing Space

The Earth's geosphere, hydrosphere and biosphere are linked into complex atmospheric activities from the troposphere, stratosphere, mesosphere, magnetosphere and ionosphere. But Elon Musk's project for thousands of Internet satellites potentially extends a form of 'metabolic rift' into the galaxy. Each electromagnetic beam from outer space purposed for fast domestic, commercial and military 5G connectivity on Earth will spread its radiation across approximately 8 kilometres of Earth's land surface. Already by 1993, Motorola was providing satellite-based accounts to some 2,000 customers around the world. Now satellite technology is being recommended for remote rural areas that cannot be provided with standard Telco services, and indeed, much of the global South might be linked into the international economy in this way. The US government has given Musk approval to launch 50,000 such satellites. Additional commercial constellations are underway with Boeing, NASA and the Pentagon; Canada's Telesat; Europe's Eutelsat; Russia's GlONASS; and China's BeiDou. Amazon also plans to offer this remote method of broadband access. Since there is an ozone-depleting effect of chlorine from satellite launcher rocket fuels, as well as acid rain and carbon soot emitted by kerosene use, Musk suggests moving to methane as satellite propellant. However, the Intergovernmental Panel on Climate Change's (IPCC) Fifth Assessment Report from 2013 says that 'methane heats the climate 28 times more than carbon dioxide when averaged over 100 years, and 84 times more when averaged over 20 years'.[46] Additionally, the functional life of a satellite is only five years, which means a continuous atmospheric traffic for replacement firing, as well as ongoing collisions of Space debris.[47]

The US military's Alaska-based High-frequency Active Auroral Research Project (HAARP) has been described as using the most powerful radio transmitter on Earth. It was intended to put out four billion milliwatts per square centimetre, but when it closed in 2006, it had just realized a quarter of this potential. As it happened, that year coincided with the collapse of bee colonies internationally. Long before this happened, ecofeminist and award-winning epidemiologist Rosalie Bertell had argued against interference with the Earth's mantle.[48] She explained that the ionosphere is what protects life from cosmic radiation, so humanly induced atmospheric destabilization can produce earthquakes and lightning strikes. While it is known that the sun bathes the Earth with ultraviolet light and X-rays, it is not generally appreciated that almost all matter, from stars to body cells, is electrically charged. The space between galaxies vibrates with subatomic particles in vast electromagnetic fields and filaments of plasma transport energy across the universe in Schumann Resonances and the van Allen Belts. Cosmic rays from outer space and radiation emitted by uranium and other radioactive elements in the Earth's crust generate the ions that carry these electric currents.[49] Against such powerful yet barely understood fields, WEF's The Great ReSet agenda projects and celebrates a 'man-made electromagnetic ecosystem' made of satellite beams and rooftop transmission towers, 5G street nodes, house wiring, WiFi routers and modems, computers, microwave ovens, cell phones and baby monitors.

Do corporate entrepreneurs have a 'right' to colonize, appropriate, extract and commodify space, given that it is a global commons? Radio frequency interference from telecommunication satellites is already hampering the observations of astronomers and meteorological forecasters in the United States, and they are calling for regulation. It is not yet known what impact satellite-directed radiofrequency beams will have on people standing on the ground in commercially activated microwave areas. Effects on biodiversity are another concern, as the foraging capacity of birds or ocean whales relies on an embodied electro-sensitivity for their navigation. The disappearance of frogs from Yosemite National Park and worldwide decline of the common house sparrow are well-documented species losses. As things stand, the impact of digitalization on climate is a blind spot in the worldwide movement of movements, and it is puzzling to find the Fourth Industrial Revolution phenomenon overlooked by environmentalists – even more so by ecosocialists. It is remarkable that a 2016 paper on the technosphere, authored by twenty-four eminent scientists for *The Anthropocene Review*, commented: 'Cataloguing and classifying its elements is a subjective and untested process, and omits certain important elements, such as *radio waves, that leave no lasting physical entity.*'[50] This claim seems astonishing coming from scholars of 'The Great Acceleration'. So also, in a public lecture on 'The Anthropocene Paradox', hosted by the University of Technology, Sydney, in May 2017, Will Steffen, an author of the above, endorsed both SDGs and geo-engineering as climate change solutions. Now, as we go to press, *The Washington Post* reports on plans to restart the problematic nuclear reactor at Three Mile Island and other suspended US facilities to meet rising energy needs caused by AI.[51]

As global capital wrestles with its problems on Earth, the patriarchal-colonial-capitalist gaze turns to extra-terrestrial resourcing; as Peter Bloom and Alberto Acosta note, 'Space represents a sort of *terra nullius* where whoever arrives first assumes sovereignty'.[52] This new frontier is opening up despite an Outer Space Treaty signed in 1967 ensuring its protection as 'common heritage of mankind'. The US military has contracted SpaceX and United Launch Alliance for scientific and military purposes. The Trump presidency promulgated a new Space Force with Starlink satellites to support an Advanced Battle Management System. The idea is to guarantee cargo or weapon delivery anywhere in the world at warp speed. The US National Aeronautics and Space Administration (NASA) has exploration agreements under its Artemis Accords with several nations; Japan and the United Arab Emirates are very interested. Bloom and Acosta note that compounding the problem of atmospheric exhaust pollution and junked satellites, space mining is seen as necessary for a net-zero carbon economy reliant on a sure supply of battery metals. The search is also on for frozen water sources to create the hydrogen to fuel galactic travel.[53]

Antonio Busalacci is well equipped to brief the public on the planetary implications of EMR technology. He presides over a University Corporation for Atmospheric Research operating under the auspices of the US National Science Foundation. In the newsletter *Corporate Risk*, Busalacci concedes that storms originating on the Sun, 'disrupt Earth's magnetic field in ways that can scramble satellite operations, distort GPS signals, knock out radio communications and power networks, and expose astronauts and airplane passengers to higher-than-normal amounts of radiation'.[54] In 2003, one large storm disabled forty satellites. Past storms have affected telegraph, power and radio. Future storms would likely knock out vulnerable electronic hardware, cell networks and routine credit transactions, affecting the life savings of ordinary folk. Busalacci's article is written for an insurance industry newsletter, but it has wider socio-ecological implications. It is significant that Lloyds of London has adopted a specific Electromagnetic Fields Exclusion Clause refusing cover for damages of this kind.[55]

Space is not empty as the 1/0 reasoning of the *Androcene* imaginary implies; nor is it simply a resource free for the taking. Life-on-Earth continues to evolve in symbiosis with the currents of great interplanetary pulses. And this is demonstrable right now, as trees scarify in massive seasonal firescapes; as insects, amphibians and mammals are disoriented.[56] It is symptomatic of the originary Humanity/Nature divide that the colonizing mindset even uses the word 'ecosystem' to describe a man-made locale. Similarly, the Chinese IT firm Alibaba refers to its algorithmic platform as an 'environmental brain'. The contemporary imperium re-enacts the 1/0 dissociation each time it responds to global warming with marketable solutions like carbon trading or end-of-pipe tech fixes. Policy becomes a matter of kicking the can down the road while persuading the public that technical experts are designing the best of all possible worlds.[57] As political leaders respond to global warming with proposals for geo-engineering the atmosphere, the approach is always already one of adapting Nature (0) to Man's convenience (1) and disposing of the rest. Given the extent to which digitalization is locked into the oil and gas industry, the notion of an 'immaterial economy' is yet another ironic reversal.

From Potsdam, the leading Anthropocene scientist Johann Rockstrom has pondered over how to make the digital revolution work for democracy. But far from the revolutionary optimism of Karl Marx's nineteenth-century class analysis, it is the sociological pessimism of Max Weber's 'iron cage' that rings true as twenty-first-century workers fall deeper into precarity. With the slump in consumer spending and state ambivalence over quantitative easing, the word 'recession' is heard again and again. IT-equipped professionals working from home pick up their own utility and insurance costs, while wealthy investors simply reshuffle their rent-generating assets. South Africans have even been pressured into micro-finance loans to digitalize their offices and homes.[58] With the business sector transitioning from manufactured 'product' to rented 'service', a weekly subscription for the use of an electric vehicle replaces car purchase. But with restructuring into the Fourth Industrial Revolution, work closures prevent the average householder from meeting mortgage commitments. Less skilled jobs disappear to automation and hygienically robotized 'reproductive care' is provided for aged and disabled bodies. While Big Pharma reduces Elder Care to the language of engineering, pandemic treatment is 'targeted' on a 'war footing' – 'ramped up' and 'rolled out' with 'shots and jabs'. Under Big Data and Big Pharma, established media institutions encounter powerful corporate censorship – albeit softened by funding, while social media infantilizes its public by scrambling news between fact, opinion and fantasy.

As political accountability is dispersed and invisibilized, social disorientation fosters moral entropy, often expressed loudly through right-wing populism and cultural resentment. In the new era of cognitive capitalism, the neoliberal university cuts funds to reflective Humanities and Social Science disciplines in favour of STEM, the Science-Technology-Engineering-Mathematics, curriculum. IT faculties introduce courses in Global Digital Property Rights (GDPR) and Citizen Management. Critical thinkers become obsolete, while technocratic operatives find meaning and reward in the design of systemic control. This sociological change is overdetermined at many levels, generated as it is by a patriarchal-colonial-capitalist culture of abstraction.

15

Digital Coloniality

Everyday Contradictions

As the United Nations process facilitates new political developments for the global ruling class, it seems quite heavily implicated in the Fourth Industrial Revolution and design of The Great ReSet. In fact, the membership list of the UN Secretary-General's High Level Panel on Digital Cooperation provides a useful survey of new twenty-first-century sociological actors. The panel comprises:

> expertise for an Electronic World Trade Platform (eWTP); a Board Member of Breakthrough Energy Ventures; a Chair of the United Arabic Emirates Council on the Fourth Industrial Revolution and Strategic Council for Artificial Intelligence Technology; a former leader of the Oslo Conservative Party; a Vice President of Google and former chair of the Internet Corporation for Assigned Names and Numbers (ICANN); a President and CEO of ICANN; a former World Bank economist now social entrepreneur; an Estonian diplomat and Chair of the Global Commission on the Stability of Cyberspace (GCSC); an Investment, Trade, and Industry Minister for Botswana – creator of a sex offender's registry; a co-founder of several Russian internet companies; a former McKinsey consultant; a former President of the Swiss Confederation; an expert in Crypto Currency and 5G strategy; a founding member of girls in ICT Rwanda; a Brazilian professor of Robotics; a director of eBay with a focus on predictive data mining; an

expert in lethal autonomous weapons systems from the Secretary-General's Advisory Board on Disarmament Matters; and the author of an academic book on Internet Governance.[1]

There is clearly a concern at the UN that women may not be willing collaborators in this new political agenda, at least judging from a 2022 UN-brokered Memorandum of Understanding (MOU) between UN Women and the banking giant Blackrock. The MOU sought women's support for UN funding by 'multilateral networking', but perhaps women's involvement was simply used to soften the fact of private sponsorship of an ostensibly neutral international agency. More 'pinkwash' than greenwash, the MOU gave women agency but at the expense of ecology, for their involvement was sought in

> a multi-stakeholder digital technology track in preparation for the 2023 Summit for the Future, to agree on a Global Digital Compact to be informed by the existing High Level Panel of Experts on Digital Cooperation, co chaired by Melinda Gates and Jack Ma.[2]

Returning to Shoshana Zuboff and her book *The Age of Surveillance Capitalism*, its subtitle reads *The Right to Sanctuary*. Zuboff is a scholar not a social radical, but the 'personal is political' passion in her work reflects both ecofeminist thinking and an indigenous 'seven generations' sensibility. To quote: 'I feel like, you know, this happened on our watch. I feel a tremendous sense of responsibility. And, this is not the world I want my kids to be living in.'[3] Among her concerns is the loss, loneliness and confusion that humans are experiencing worldwide: as a species evolved for conviviality is locked into a numb hyperconnected electronic future. Young people, though critical of the current political system, may still be addicted to cell phones, Instagram or Netflix. Riding the bus or clustered in coffee shops, they are forever comparing selfie shots as if only their mirror image by 'the device' affirms their worth. But modern narcissism costs the Earth. And when everyday experience is reduced to a repetitive circuit of hand-keyboard-eye-screen-hand-keyboard, then sensory deprivation and ethical disorientation set in. Many youth grasp the historical contradiction imposed on them by digital technologies and even try to defuse their cognitive dissonance with the playful acronym WEIRD – Western, educated, industrialized, rich and democratic. Most agree that artificial intelligence (AI) is not value free, with algorithm designers often having racist, sexist and ageist attitudes.[4] That said, South Africa's president Ramaphosa called for an AI Forum in his capacity as chair of the African Union, and the University of Johannesburg vice chancellor serves on a Commission for the Fourth Industrial Revolution. Meanwhile, as the Centre for Integrated Post-School Education and Training at Nelson Mandela University noted recently, 800 million workers globally will have lost their jobs to robots by 2030.[5] Does anyone seriously believe this press-button 'development model' can liberate the Global South? Notwithstanding such concerns, some indigenous scholars do seek to take advantage of the *1/0 imaginary* in ways that might serve their traditions. In reaction to old Eurocentric rhetorics used to demean and control peoples as 'primitive', they argue that software can be made to talk to hardware in ways that respect indigeneity. For example, computer documentation might preserve Anishinaabe creation stories; programmes might be adapted to revitalize indigenous languages; block chain might be used by indigenous business ventures to build self-determination.

In 2020, an *Indigenous Protocol and Artificial Intelligence Position Paper* was released in the United States.[6] Two lead authors, Jason Lewis and Noelani Arista, professors in Canada and Honolulu, respectively, describe themselves as Kanaka Maoli. Other contributors to the *Position Paper* have Australian, New Zealand and North American backgrounds, and their skills range through neuroscience, robotics programming, art and film. The vision is that indigenous onto-epistemologies may reshape modern technologies in ways that reflect the values of locality, relationality and reciprocity. Lewis, whose field is Computational Arts, is insightful when he writes:

> For many of us working in or with the experience of the high-tech industry, it was a relief to focus on such questions rather than the tired tropes of a technology elite that recursively chases its own tail upon a ground of epistemological blindness, cultural prejudice and myopic misanthropy . . . Asking our questions asserts our sovereignty, over our minds, our lives and our futures.[7]

The *Protocol* project has parallel initiatives with public-private collaborations issuing manifestos on machine ethics in technology design – the Montreal Declaration is one example. Professional bodies like the Institute of Electrical and Electronics Engineers have also adopted design guidelines to protect 'well-being'. But the indigenous authors of the *Position Paper* point out that these statements remain anthropocentric – based on the 'prim logic of linearity' and the colonial mindset as a 'philosophic monoculture'.[8] One clear failing in machine design standards is the lack of recognition capacity for Black faces. In short, 'Computer science has a major White guy problem'.[9]

Modernity

The rationale for the *Indigenous Protocol and Artificial Intelligence Position Paper* uses the posthumanist vocabulary currently popular in the Humanities and Social Sciences. This follows philosopher Bruno Latour and more recently Donna Haraway's paper *Making Kin in the Chthulucene.*[10] As the *Protocol* introduction reads:

> Man is neither height nor centre of creation. This belief is core to many Indigenous epistemologies. It underpins ways of knowing and speaking that acknowledge kinship networks that extend to animal and plant, wind and rock, mountain and ocean. Indigenous communities worldwide have retained the languages and protocols that enable us to engage in dialogue with our non-human kin, creating mutually intelligible discourses across differences in material, vibrancy, and genealogy.[11]

Here the object world – natural and manufactured – is grasped respectfully. Kekuhi Kealiikanakaoleohailiani, one of the workshop participants, concludes that AI 'is a sort of extension of ourselves. I don't think it's any different from having created a canoe.'[12] Equally empathic is the Latour-inspired suggestion that 'Giving thanks to *a system* is the same as giving thanks to a relative'.[13] The indigenous design ethic will be about 'entwining trust and care with AI'. When regarded as 'kin',

AIs have needs, just as humans do. They need clean and nourishing food (a data diet), security, comfort in temperature, and capacity for fulfillment ... and while data seems sterile, placeless, quantifiable, and scientific, it is entwined with place-based knowledge whether it is cultivated on land or in territories of cyberspace.[14]

While emphasizing that there is not one indigeneity, the *Position Paper* blends quantum theory as interpreted by Karen Barad with indigenous understandings of how the world works.[15] Thus,

The Anassin AI understands the universe as flow, constellations of forces contracting and relaxing to form the always-becoming-always-unravelling knots of Newtonian causality with which we consciously interact. It sees past-present-future as a unified whole, a four-dimensional volume where everything that has occurred, is occurring and will occur – one just has to know the coordinates.[16]

The Fourth Industrial Revolution is defined as 'foundational' here, with computing on par with railroads and electricity, but when the *Position Paper* takes up an embedded ethic, it begins to uncover some material limits. For example, in considering the principle of Respect for Territory, how does an indigenously designed AI justify the destructive mining of rare minerals needed for making machine components – or the mega-global energy burn that comes with cloud computing and the Internet of Things? There is also worker exploitation and toxic exposure to bodies on the factory assembly line; water and soil contamination from IT waste disposal.[17] The *Position Paper* acknowledges that 'Robust ethics are necessary for . . . extracting natural resources on a global scale. International and domestic regulation is necessary, as well as a movement to produce recyclable materials that can build computers, as well as the right to repair.'[18] This response follows the ecomodernist optimism favoured by business and the international technocrat class. As they argue, industry regulation can ease rates of planetary damage, while engineering innovation 'will dematerialise' energy and resource usage.[19] However, a closer reading of the environmental literature shows that neither regulation nor dematerialization is effective in reducing human impacts on climate, biodiversity or health. It is all but impossible to reconcile imposts of the Fourth Industrial Revolution with a precautionary 'seven generations' ethic. Statements like 'Can we infuse a beloved tree with the technology to "tell" us how it feels?' seem to imply some kind of mechanized interface, which is surely a violation of the plant.[20]

The *Position Paper* is an elegant document but it does marginalize fine-tuned indigenous understandings of habitat. In fact, the high modern posthumanist desire to leave behind oppressive categorical boundaries tends to blur critical discernment. It is one thing to refuse the traditional Eurocentric Humanity/Nature dualism by endorsing human care for other species; it is another to suggest that inert objects like rocks or robots are 'kin' with capacity for agency and reciprocity. A flowing river or for that matter a technology may produce an 'affect' in people, but this is not 'affection'. An 'effect' is qualitatively different from an 'affect', which is subjectively intentional. For sure, dualisms, such as Humanity over Nature, Man over Woman, White over Black, Mind over Body, are ideological fabrications used to diminish the second dualist term. But these opposites should be taken apart and exposed as politically hegemonic of the *Androcene* era; they cannot simply be 'essentialized' or cancelled in a one-dimensional way. Besides this, as anthropologist Alf Hornborg argues, people need to hold on to some analytical binaries simply in order to distinguish 'this' from 'that'. The prevailing intellectual hesitancy to concede categorical differences becomes nonsensical when it confuses ideological categories with cognitive processes.[21] If the epistemological distinction between human and nature, or subject and object, lapses altogether, then almost anything can be claimed – convenient, perhaps, for an emerging authoritarian society running on fast media and fake news.

Data Sovereignty

The story of electoral manipulation using Facebook profiles harvested by Cambridge Analytica is well known. But when digitalization becomes the basis of everyday life it can exacerbate power imbalances of class, ethnicity, sex/gender and age in many ways. Given the ubiquity of internet data mining the *Indigenous Protocol and Artificial Intelligence Position Paper* expresses concern over how indigenous information can be protected. The question of what eventually happens to genetic test results from companies like Ancestry .com is a significant one for dislocated or threatened minority populations. While digital technique can help find family members scattered by invasion, it also repackages that indigenous data under an alien copyright law.[22] As the authors put it:

> Indigenous people must control how their data is solicited, collected, analysed, and operationalised. They should decide when to protect it and when to share it, where the cultural and intellectual property rights reside, to whom those rights adhere, and how these rights are governed.[23]

While nothing will take away the scars of colonial suppression, indigenous self-knowledge goes a long way in healing the trauma of broken ancestry.[24] Can the 'chant, prayer, law, history, story' stored in computer memory bring meaningful immersion? Or are there more vital paths to cultural reclamation? In Australia since 2019, the UN Year of Indigenous Languages, First Nation peoples have designed a public education programme called Word-Up to share some of the 1,000 tongues spoken on the continent over the past 60,000 years.[25] That community relation to language is embedded in peoples' cultural landscapes known as Country; so today, Black and white folk go out together across the island continent, to learn how to gather bush foods and undertake customary land maintenance.

Yet cross-cultural communication can always be marred by tensions and ambiguities. In the case of the Canadian *Position Paper*, its contributors are creative, highly qualified and enthusiastic about AI careers. Their text resonates with the UN approach to Indigenous Governance and Traditional Knowledge Systems. It also echoes 1990s moves by the UN to encourage indigenous involvement in genetic engineering, with the promise of economic benefits to be gleaned from patenting traditional biodiversity. Decades later, the new commodity is not only genetic codes but also value in information coding across the board. The narrative of 'shared benefits' readily resonates with the think tanks and hidden persuaders of the imperium. Among these is California's business-oriented Breakthrough Institute, onetime host to

Latour. However, as noted, the dispersal of intentional agency into the object world flattens the focus on hierarchies of power embedded historically in the patriarchal-colonial-capitalist system. In this ecomodernist idealism, ontological relations are neutralized as 'semiotic' rather than materially lived.

There are vast profits to be had from the Fourth Industrial Revolution, especially now that cybersemiotics and data gleaning are pathways to a Smart Future. The Covid-19 pandemic opened into a full spectrum of opportunities for AI applications. A motivational speech to the corporate sector by the World Economic Forum (WEF) leader Klaus Schwab is telling, in this respect:

> We stand on the brink of a technological revolution that will fundamentally alter the way we live, work, and relate to one another. In its scale, scope, and complexity, the transformation will be unlike anything humankind has experienced before. We do not yet know how it will unfold, but one thing is clear: the response to it must be integrated and comprehensive, involving all stakeholders of the global polity, from the public and private sectors to academia and civil society.[26]

Does this prefigure the posthuman Singularity: a world in which the 1/0 consciousness absorbs all materiality into itself, so all that remains is the sphere of 1?[27]

What WEF calls The Great ReSet merges governments, Big Data, weapons research, Telcos, agri-tech, the Internet of Things and Internet of Bodies. Ideally, a new *Position Paper* would address this twenty-first-century will to power. However, as things stand, the *Protocol* only acknowledges the financial support of the Social Sciences and Humanities Research Council of Canada (SSHRC) and input from the Canadian Institute for Advanced Research (CIFAR). The latter is described as a global charitable organization, whose remit 'convenes extraordinary minds to address the most important questions facing science and humanity'.[28] CIFAR is internationally partnered by France's Centre National de la Recherche Scientifique (CNRS) and by UK Research and Innovation (UKRI). As the authors explain: 'The AI & Society Program is the fourth pillar of the CIFAR Pan-Canadian AI Study, a $125-million investment from the Government of Canada, with the goal of supporting Canada's leadership in machine learning and training. It is also supported by Facebook and the RBC Foundation.'[29]

Alongside these arguments for economic growth, Maori contributing author to the *Position Paper*, Professor Hemi Whaanga from the University of Waikato, offers a cautionary note about cultural assimilation. Quoting his colleague Professor Rangi Matamua, he asks, is AI 'A New (R)Evolution or the New Colonizer for Indigenous Peoples?'[30] To control a people's culture is to control their tools of self-definition. As Whaanga notes, the power of indigenous knowledge systems is embedded in land; in other words these cultures do not emerge from the dissociated psyche with its Humanity/Nature divide but are created in materially embodied interaction with the metabolism of Earth processes. By contrast, posthumanist methodologies are purely cognitive, even while brandishing the language of bodily 'affect'. More than this, their relativism means there is no basis for contesting abuses of power. It is disturbing to find decolonial activists adopting such a discourse.

Double Binds

The stages of colonization are well worn: invasion, conquest, appropriation, occupation, suppression and 'justification'. And not to forget that the terrain of politics includes our very body cells – the sickness of diabetes a signifier of colonization *par excellence*. Looking at projects for indigenous AI, it seems that the logic of 'colonial justification' has deleted historical memory. Certainly there is a sense in Aotearoa that the Fourth Industrial Revolution is universalizing and epistemicidal, if not potentially genocidal.[31] It is good to see that coordinators of the Canadian *Position Paper* coordinators share this heartfelt criticism upfront in their report. But then again, the *Androcene* imperium preserves its power by keeping peoples in conflict among themselves. The very word 'stakeholders' encapsulates the 'divide and rule' psychology of governments. It is even seen in response to ecofeminist criticism of cyberculture, when the IT industry cultivates xenofeminists and Geek Girls to embrace coding and hackathons, 3D printing and wearables, game design and start-ups.[32] The parallel subsumption of indigeneity has been a predictable development – even though youth is the main target of twenty-first-century marketeers.

Is there any hope for global change when social justice movements themselves become embedded in a network of globally militarized AI technologies? When political activists depend on digital networks for sharing ideas and organizing, they are/we are skewered by a profound material contradiction. The bind is especially sharp for environmentalists, recalling that:

> To make just one solar panel generates up to 70 kg of carbon dioxide . . . The expected increase in the number of solar panels will generate 2.7 billion tons of carbon emissions – the same as 600,000 vehicles on the road in a year.
>
> An email with an attachment uses as much electricity as a high-wattage energy-saving light bulb does in one hour. Every hour, ten billion emails are sent – equivalent output of 15 nuclear power stations in an hour.
>
> The information and communication technology sector consumes 10 per cent of the world's electricity, producing 50 per cent more greenhouse gases than air transport in a year.[33]

African scholar Sabelo Ndlovu-Gatsheni expresses concern over of the transition to violence that follows from mapping and coding; and indeed, it is a form of *terra nullius*.[34]

Arturo Escobar finds most NGOs operating on the received logic of Eurocentric modernity – its markets and its universals. Together with South American feminist Marisol de la Cadena, he emphasizes how critical it is to respect and learn from the relational 'lived values' of indigenous peoples. Escobar wants activists to ask

> What habitual forms of knowing, being, and doing does a given strategy contribute to challenge, destabilize, or transform? For instance, does the strategy or practice in question help us in the journey of de-individualization and toward recommunalization? . . . Does it make us more responsive to the notions of multiple reals and a world where many worlds fit? Does this shift encourage us to entertain other notions of the possible, significantly different from those on offer by capitalism, the state, the media, and most expert institutions? To what extent do our efforts to depatriarchalize and

decolonize society move along the lines of liberating the Earth and weaving the pluriverse effectively with others, human and not?[35]

From this decolonial vantage point, it is useful to turn back to sustainability strategy.

Decoupling?

Is it really possible to address the scale of twenty-first-century planetary crises by recycling, renegotiating trade, taxing business or putting industry in public hands? On a thermodynamic level, the full materiality of matter-energy cycles is rarely factored in by experts or conceded by government. Consider the production/consumption chain:

> Nature – Extraction – transport – Smelting – transport – Manufacture – transport – Marketing – transport – Consumption – transport – Repair – transport – Wastepit – Nature

Add to this the enclosure of community lands for roads and transmission lines. Even renewable-energy installations must be manufactured 'somewhere' and rely on batteries for power storage, devices that must be disposed of 'somewhere else'. Other would be climate savers like electric cars pull in ambient electromagnetic radiation from 5G street chargers, while lithium batteries are prone to self-ignite.[36] Decoupling resource use from ecological damage may be demonstrable for a single technology; but once production involves complex global transactions the effect is annulled. And where is the assessment of biomedical impacts when regular working families are using asbestos, CFCs, chlorine, dental mercury, lead paint, nanotech, PCBs, pesticides, PFAS, plastics, synthetic oestrogens and toxic dyes? At the same time, these are threats to animal, fish and plant species.[37] Government-administered Auditing and Certification, Monitoring and Evaluation is pointless if the commodity itself is harmful.

Capitalist consumer economies serve the desires of 20 per cent of the world population but simply degrade daily living conditions for the rest. As discussed already, environmental remedies like Climate-Smart Agriculture,

Geo-Engineering or Earth System Governance are politically compromised too.[38] Meanwhile, the circular economy notion and the Green New Deal are solutions that stand for many things. They would adjust 'the ecological footprint' by reducing material input and waste output with closed-loop production; but like 'the principle of sustainability', the circular economy can be taken over by productivist reasoning – as seen when systems engineers talk of 'technical nutrients'! The UNEP *Global Green New Deal* has the advantage of being international in scope, but it is still framed by the master lens. Thus, with the UN increasingly dependent on corporate moneys to meet its agenda – the triple bottom line of planet, profits and people – it approaches ecology in ways that make no thermodynamic sense. In the Global North, there are Right-leaning Green New Deal proposals from the World Economic Forum and the European Union, centrist versions from US Democrats and United Kingdom Labour and Leftist accounts from DiEM25 and DSA-USA, but such reform programmes appeal mainly to educated professionals reliant on urban lifestyles. In this context, a Just Transition usually means 'internalizing' business-generated externalities by remaking trade agreements around climate standards and labour rights. The Breakthrough Institute, a think tank founded by pollster Ted Nordhaus and lobbyist Michael Shellenberger, has added to public confusion on these matters with an *Ecomodernist Manifesto*, co-authored with Stewart Brand. This advocates 'decoupling' through urban density planning, desalinization, transgenic agriculture, hydro dams, nuclear fusion and geoengineering.[39] Some of their policy options are said to be 'transitional' – like 'absolute dematerialization', for example, whereby external impacts are said to rise quickly, then decline as the economy grows. However, degrowth economists Jason Hickel and Giorgos Kallis judge absolute decoupling to be illusory.[40]

Is post-industrial dematerialization any more than ideology? Vijay Kolinjivadi and Ashish Kothari write that almost nothing illustrates the need to decolonize ecology so clearly as the Green New Deal idea in the Global North. All it does is 'spur the imperialist quest for cheaper resources and labour' to satisfy 'ecofriendly' ambitions.[41] Devised in affluence, popular new deals perpetuate the linear reasoning of a patriarchal-colonial-capitalism that equates industrialization with 'progress' as universal prerequisite of 'the good life'. Overlooked is the fact that everyday consumption in the Global North means land grabs for biofuels, coal or hydro projects and mining of rare earths, as men, women and children of the Global South work in poisonous environments to provide 'clean and efficient' products for the Global North. To reiterate the famous 'imperial mode of living' formula of Uli Brand and Markus Wissen: for every kilogram of purchased product in Germany, 5 kilos is removed from someone else's land. This goes unquestioned by everyday citizens, middle and working class alike.[42] UN rhetorics notwithstanding, colonization continues in the twenty-first century and it still rests on cost shifting from privileged to marginalized, often racialized, peoples. Even then, resource grabs are just one level of dispossession in the hierarchy of debt that keeps the modern global system afloat. The 2030 Agenda and Sustainable Development Goals, as scrutinized by Noam Chomsky, Naomi Klein and Medha Patkar, show the same limitations. The bottom line is that three or more Earths would be needed to democratize an ecomodernizing industrial lifestyle across the planet. The circular economy is credited with increasing job opportunities, but what is 'a green job' really? The one-size-fits-all thinking behind prevailing international policy does not acknowledge the many thousands of green jobs already existing in the Global South. Here, a living circular economy can be found, which has regenerated people and nature together over centuries.[43] But as agency-funded development projects cut across this integral society-nature metabolism, valuable ecological labour skills are lost. There is a critical difference between the hegemon's quantitative notion of productivity and a qualitative vision grounded in practical meta-industrial knowledge of nature. So too, low-impact, self-sufficient farm communities have the advantage of being buffered from the ups and downs of global capitalist crises.

16

Land Is Law/Lore

Another Ontology

In the steady rise of the patriarchal-colonial-capitalist continuum, the evolution from androcentric contests over resources like women's bodies and land led to more geographically ambitious appropriations, which, in turn, fuelled the modern money economy. Along the way, the nation state model was adopted across the globe as the legal framework for 'development'. Modelled on the seventeenth-century Treaty of Westphalia, which ended a thirty-year war among European tribal interests, arbitrary colonial boundaries were drawn across living ecologies and ethnicities, imposing a false political unity under the banner of 'nationhood'. The twentieth-century *Androcene* was consolidated by capitalist opportunism, as a post-war Bretton-Woods Agreement established the global chaos of neoliberal trade. Dipak Gyawali reflects on how

> the main task of nation-states is 'bordering, ordering and othering'. It is in this process of 'ordering' by governments that the 'othering' occurs, leaving communities marginalized and vulnerable, as well as environment such as rivers, forests and wildlife desecrated.[1]

The colonial experience is clear as Vanuatu villagers grieve the loss of their Oceania *kastom ekonomi*:

Foreigners used to tell us that we needed to Change; then they told us we needed Progress, and now they tell us we need Development. It usually means they are after something we have – either our forests or our land or what is under our land, or our souls, or language or culture, or our feeling of contentment with our way of life . . .

[But] smart Melanesians tend to see *Kastom* as protecting them from bad development and the disease that comes with it – *Sick blong mane* – or money addiction.[2]

EcoSufficient Ethics

Today, yet another biopolitical reset is underway, with populations redefined as 'assets' and a controlled social demolition assisted by the global pandemic. The new capitalism is proudly posthumanist, and as Big Data surveillance maps the public sphere, Big Pharma plans a comprehensive WHO Treaty that would bring even the governments of nation states under its direction in the event of future viral outbreaks. Facing a conjuncture of crises, workers, indigenous Peoples', women and youth call out this latest *terra nullius,* even while heavily preoccupied over climate and seeking a Peoples' Tribunal on nature's behalf.[3] Too often, well-intentioned actions for change are turned around to reinforce the *sickness blong mane* and other universalizing fetishes of modernity. This has happened with proposals for circular economies, smart agriculture, Earth Governance, genetech and SDGs. Yet against the Eurocentric *cogito* – 'I dominate, therefore I am' – consider the South African *ubuntu* ethic or *susu* from Ghana, premised on the conviction that 'I am because you are'. Similar ways of worlding that understand humans-as-nature-in-embodied-form are *buen vivir*, ecovillages, *hurai*, the gift economy, *kyosei, sentipensar* and *swaraj*. These political ontologies are stirring interest across the globe.[4] The ideal of food sovereignty is fundamental to this local but pluriversal sensibility – not to be confused with the UN Food and Agriculture Organization's idea of 'food security' designed for agro-industrial profiteers. The familiar costs of that are petro-farming, terminator seed, polluting transcontinental trade and lifeless processed foods.⁵ To recall Vandana Shiva's words: the ultimate achievement of capitalist farming is the Dust Bowl. Sadly, international awareness of 'externalities' like these is uneven. Affluent metropolitan nation states, and even elites from the geographic periphery, downplay, even deny, the damage of their lifestyles on nature's metabolism. Academic analyses too, may be compromised by the neoliberal trend to corporate funding of research institutes. But as long as entrepreneurs chase growth, inside and out of UN agencies, the promise of a fair and sustainable distribution of 'the social product' is simply nonsensical when the world economy already overshoots the Earth's capacities each year.⁶

The first step towards a transversal politics is to resist the ancient dissociation of Humanity from Nature, Production from Reproduction. For an *Androcene* that severs its own embodiment becomes a necrocene and must satisfy its lack with signifiers like Religion, Money, State and, soon enough, The Singularity. On the other hand, indigenous peoples know well that energy flows within human bodies are the same energies that animate the Earth. This perception relies on an epistemological fusion of self-with-field, and across the world, people who work regeneratively with land, animals and children are very sensitive to the material timings of living cycles. This way of worlding, found at both the domestic and geographic margins of capitalism, puts basic needs and everyday 'use value before exchange value'.⁷ The problem is that at the UN and multilateral summits, the experience of caregivers or small farmers is treated as 'cultural' not economic, devaluing their skills and keeping their vernacular science outside the discourse of governance. That said, since the 1999 Seattle Peoples' Caucus attempts to revitalize decolonial politics have been initiated by organizations such as the World Social Forum, Systemic Alternatives and Global University for Sustainability. Rather than argue for Basic Income handouts from a monetized nation state, the alter-globalization movement prioritizes eco-sufficient local provisioning, grounded in time-tested communal traditions of direct democracy. Further radical leadership comes from advocates of solidarity economies, permaculture, *buen vivir* and commoning.⁸

As the late David Graeber so famously said: workers have had enough of routinized, meaningless, 'bullshit jobs' in a world system based on the glossy marketing of 'false needs'.[9] But from Catalonia, via Hungary, to Greece, and beyond, theorists and activists are now promoting an economics of degrowth.[10] Radical versions of degrowth converge with the Earth-based 'subsistence model' advicated by ecofeminists Maria Mies and Veronika Bennholdt-Thomsen three decades ago. Thus,

- Regional economies respecting the living limits of nature
- Land held in common for the direct satisfaction of needs
- Reciprocity of rural and urban, producer and consumer
- Food sovereignty, not supply chains and dumped surpluses
- Local trade to protect creativity and reduce global pollution
- Easy-to-use technologies crafted to avoid ecological damage
- Money as a means of circulation, not means of accumulation
- No sexual division of labour, all doing reproductive work.[11]

Note that ecofeminism is not an 'identity politics' that argues from women's victimhood, nor is it based on some essentialist patriarchal myth of an 'innate feminine morality'. Likewise the contemporary appraisals of 'indigeneity' are based on a materialist ontology, not some eurocentric idealization of 'the noble savage'. Unsympathetic, dualist readings such as these, simply reveal old patriarchal-colonial-capitalist defence mechanisms at work. The decolonial argument is that people realise their full species powers in the labour of nurturing the humanity-nature metabolism – a point the young Marx himself appreciated.[12] This embodied materialist sensibility graces the peasant politics of *Via Campesina*, Uruguay's World Rainforest Movement, the initiatives of Korean Gabaewul seed savers, panAfrica's WoMin, Venezuela's Chavistas, Deccan-based Dalit Women and the *jin jiyan azadi* re/sisters of Rojava. The World March of Women and PeaceWomen Across the Globe reinforce these efforts for international justice and peace. The call is to divest from the egoism of war and corporate competition: to rematriate 'territories of life' and rebuild community from the ground up.[13]

Bioregional Politics

On the island continent of Australia, the Kombumerri philosopher Mary Graham and the Mardoowarra activist Anne Poelina teach how First Nation peoples prioritize an ethic of 'place'. As they say, living in the Country of one's ancestors grows a sense of self and belonging. Where Land is Law/Lore, people have 'an ontological compass'.[14] Local grassroots projects for re-worlding join up indigenous science for grounded sustainability, subsistence economies for nurturing equality and care; and advance a global politics where wealth is found in cultural diversity. Across continents, as peoples' networks share the joy of this commoning, they resist the new face of *terra nullius* while reweaving the thread of Life-on-Earth. In Pune, India, Kalpavriksh activists are fostering cross-continent conversations known as the Global Tapestry of Alternatives. Other transversal political moves include Territories of Life groups enabling bioregional self-reliance and Bhutan's inspired commission for Gross National Happiness convened by local Peoples' Assemblies. Here, well-being is understood as an integration of ecological, biological, psychological and social energies.[15] Alter-globalization activists in the global North are learning quickly from organisers in the global South. For example, in the United States, Jackie Smith, a long-standing world-system scholar, continues her World Social Forum activism by fostering a transformative movement politics within urban livelihoods.[16] In the EU, the Left network Progressive International has conservation principles highlighting – Conviviality – Diversity – Decommodification – Valuing the Sacred – Decolonization – Social Justice – Direct Democracy – Redistribution – Subsidiarity – Global Connections – Resistance – Redefining Power.[17] The challenge remaining for many Left groupings, however, is where they stand vis-à-vis the role of the state as *Androcene* incarnate.

Will young people today let themselves be encircled by Credit – Computers – Cars – Climate – Covid – and Cognitive Capture? No. In the United States, a nationwide body of students known as Dissenters is protesting university ties with the Pentagon and its private contractors.[18] And while governments and multilateral agencies ignore the Rights of the Child, high school kids in Germany challenge political myopia with Fridays for Future. So too, the energies of Extinction Rebellion and *Ende Gelande* enliven an Earth Democracy guided by interdependence and flow. Then, in February 2024, guided by the China-based Global University for Sustainability, the dream that Another World is Possible! was re-ignited. Now, 50,000 grassroots activists from some 1,400 organizations travelled to a newly convened World Social Forum in Kathmandu. As well as climate, sustainability and systemic discrimination of 'minority' groups, the Nepal workshops covered global finance and power elites, military spending and austerity, repression and securitization. With WSF emphasizing local sufficiency, community assemblies and stewardship for future generations, the 'bioregional' path to restoring Humanity-Nature relations was explored with enthusiasm.

Bioregionalism is about 'ensuring that political boundaries reflect ecological and cultural ones'.[19] In honouring this, people move from the androcentric to an ecocentric world view. But the question is: How to de-link from the historical artifice of global nation states and corporations, so as to redraw life-responsive political boundaries? Can publics and politicians be motivated to ask:

How coherent are the routine sustainability proposals of 'governance as usual'?

Whose lives are tacitly subsumed and consumed by ecomodernist development?

Would a 'bioregional deal' avoid the matrix of social, embodied, generational and species debt?

How can eco-sufficient provisioning become a reality for everyone who seeks life-affirming change?

Should we set up a shadow Peoples' Security Council and dual-power the *1/0 imaginary* of the UN and its patriarchal-colonial-capitalist agencies?

By Androcene reasoning, both Right and Left politics assume that labour must be 'productive', which is a way of saying that it is about transforming material nature into something else – man-made – and thus having 'value'. Too often, even progressive political economists, Greens and ecosocialists, will argue quantitatively by trying to relocate care work or organic farming – 'holding labours' – inside this abstract formal economy. Yet there is another labour and 'another value', experienced directly as natural ecosystems flourish and human bodies with them. This reproductivity does not need to be measured, its value is simply felt as children dance by, as trees bear down with fruit, as corn shoots up from the soil. People enjoy this wealth when clean river waters deliver new season's fish. Holding labour is 'a way of worlding' that is learned when the senses open to another ontology. Karin Amimoto Ingersoll recognizes this ecocentric sensibility among the fishermen of Hawaii:

> a non-instrumental navigational knowledge about the ocean, wind, tides, currents, sand, seaweed, fish, birds, and celestial bodies, as an interconnected system that allows for a distinct way of moving through the world. In this oceanic literacy, the body and the seascape interact in a complex discourse . . . an alternative to the grand narrative of Western thought-worlds, which keep our 'selves' separate . . .
>
> Seeing thus becomes a political process . . . a reading of all memories and knowledges learned within oceanic time and space but which have been effaced by rigid colonial constructions of identity, place, and power . . .
>
> [Too] much of the world proceeds without memory, as if the spaces we inhabit are blank geographies, and thus available for consumption and development . . .[20]

The ongoing dissociations of the patriarchal-colonial-capitalist era are sending our efforts to protect Life-on-Earth askew. Can critical responses to modernity like ecosocialism or ecofeminism escape that historical *cul de sac*? The answer to this question begs a further volume or two.

Notes

Preface to the Trilogy

1 Mark Furlong, 'Looking for Eros in the Long Hard Rain of Climate Collapse', *Arena*, 13 (2023) 74–7.

2 Andrea Saltelli and Monica Di Fiore (eds.), *The Politics of Modelling: Numbers between Science and Policy*. Oxford: OUP Academic, 2023; Ariel Salleh, Mary Mellor, Katharine Farrell, and Vandana Shiva, 'How Ecofeminists Use Complexity in Ecological Economics' in K. Farrell, T. Luzzati, and S. van den Hove (eds.), *Beyond Reductionism*. London: Routledge, 2012.

Chapter 1

1 Nafeez Ahmed, 'U.S. Military Could Collapse Within 20 Years Due to Climate Change, Report Commissioned by Pentagon Says', *Motherboard Tech by Vice*, 24 October 2019: https://www.vice.com/en_us/article/mbmkz8/us-military-could-coll.. .years-due-to-climate-change-report-commissioned-by-pentagon-says.

2 Barry Gills, 'Deep Restoration: From the Great Implosion to the Great Awakening', *Globalizations*, 17 (2020) 577–9.

3 Editorial, 'Agrarian South Network Platform for Dialogue', July 2020: www .agrariansouth.org/new.

4 Paul Crutzen, 'The Geology of Mankind', *Nature*, 415 (2002) 23; Lewis Dartnell, *Origins: How the Earth Made Us*. London: Penguin, 2018.

5 William Ruddiman, 'The Greenhouse Era Began Thousands of Years Ago', *Climatic Change*, 61 (2003) 261–93; William Ruddiman, *Plows, Plagues, and Petroleum: How Humans Took Control of Climate*. Princeton, NJ: Princeton University Press, 2005.

6 Will Steffen, Jacques Grinevald, Paul Crutzen, and John McNeill, 'The Anthropocene: Conceptual and Historical Perspectives', *Philosophical Transactions of the Royal Society A*, 369 (2011) 842–67.

7   Johann Rockström, W. Steffen, K. Noone, Å. Persson, F. Chapin, E.Lambin, T. Lenton, M. Scheffer, C. Folke, H. Schellnhuber, B. Nykvist, C. de Wit, T. Hughes, S. van der Leeuw, H. Rodhe, S. Sörlin, P. Snyder, R. Costanza, U. Svedin, M. Falkenmark, L. Karlberg, R. Corell, V. Fabry, J. Hansen, B. Walker, D. Liverman, K. Richardson, P. Crutzen, and J.A. Foley, 'A Safe Operating Space for Humanity', *Nature*, 461 (2009) 472–5.

8   Noel Castree, W. Adams, J. Barry, D. Brockington, B. Buscher, E. Cornera, D. Demeritt, R. Duffy, U. Felt, K. Neves, P. Newell, L. Pellizzoni, Kate Rigby, P. Robbins, L. Robin, D. Bird Rose, A. Ross, D. Schlosberg, Z. Sorlin, P. West, M, Whitehead, and B. Wynne, 'Changing the Intellectual Climate', *Nature Climate Change*, 4 (2014) 763–8.

9   Steffen et al., 2011, p. 853.

10 Andreas Malm and Alf Hornborg, 'The Geology of Mankind? A Critique of the Anthropocene Narrative', *The Anthropocene Review*, 1 (2014) 62–9; Jason Moore, *Anthropocene or Capitalocene? Nature, History and the Crisis of Capitalism*. Oakland, CA: PM Press, 2016.

11 Eric Swyngedouw and Henrik Ernston, 'Interrupting the Anthropo-obScene: Immuno-Biopolitics and Depoliticization in the Anthropocene', *Theory, Culture, and Society*, 35 (2018) 3–30, p.1.

12 John Clark, *Between Earth and Empire: From the Necrocene to the Beloved Community*. Berkeley, CA: PM Press, 2019.

13 John Bellamy Foster, *Capitalism in the Anthropocene: Ecological Ruin or Ecological Revolution*. New York: Monthly Review Press, 2022, pp. 472–3.

14 Kohei Saito, *Marx in the Anthropocene: Towards the Idea of Degrowth Communism*. Cambridge: Cambridge University Press, 2022.

15 Bruno Latour, *Politics of Nature: How to Bring the Sciences into Democracy*. Cambridge, MA: Harvard University Press, 2004; Donna Haraway, 'Anthropocene, Capitalocene, Plantationocene and Chthulucene: Making Kin', *Environmental Humanities*, 6 (2015) 159–65; Richard Grusin (ed.), *Anthropocene Feminism*. Minneapolis: Minnesota University Press, 2017.

16 Hamed Hosseini, 'Transversality in Diversity: Experiencing Networks of Confusion and Convergence in the World Social Forum', *International and Multidisciplinary Journal of Social Sciences*, 4 (2015) 54–87.

17 Ariel Salleh, *Ecofeminism as Politics: Nature, Marx, and the Postmodern*. London: Zed Books, 1997/2017.

18 Franciszek Chwalczyk, 'Around the Anthropocene in Eighty Names: Considering the Urbanocene Proposition', *Sustainability*, 12 (2020) 4458–501.

19 Bell Hooks, *Feminist Theory: From Margin to Centre*. Boston: South End Press, 1984; Bell Hooks, *The Will to Change: Men, Masculinity, and Love*. New York: Atria Books, 2004.

20 Walter Mignolo, 'The North of the South and the West of the East: A Provocation to the Question', 2014: www.ibraaz.org/essays/108.

21 Gianmaria Colpani, 'Crossfire: Postcolonial Theory between Marxist and Decolonial Critiques', *Postcolonial Studies*, 25 (2022) 54–72.

22 John Clark, 'Ecofeminism', *Encyclopedia of Applied Ethics*, 2021: http://www.sciencedirect.com/topics/social-sciences/ecofeminism; Sandra Harding, *Is Science Multi-Cultural? Potcolonialisms, Feminisms, Epistemologies*. Bloomington, IN. Indiana University Press, 1998; Wendy Harcourt, *Feminist Perspectives on Sustainable Development*. London: Zed Books, 1994.

23 Arturo Escobar, *Designs for the Pluriverse: Radical Interdependence, Autonomy and the Making of Worlds*. Durham, NC: Duke University Press, 2019, p. 27.

24 Leonardo Figueroa-Helland, 'Indigenous Pathways Beyond the Anthropocene', *New York University Environmental Law Journal*, 30 (2022) 347–412.

25 Luigi Pellizzoni, *Ontological Politics in a Disposable World: The New Mastery of Nature*. London: Ashgate, 2015; Robbie Shillian, 'Decolonizing Politics', 2021: https://blogs.lse.ac.uk/lsereviewofbooks/2021/08/24/long-read-review-decolonizing-politics-an-introduction-by-robbie-shilliam/

26 Markus Kroger, *Extractivisms, Existences, and Extinctions: Monoculture Plantations and Amazon Deforestation*. London: Routledge, 2022, p. 83; Extinction Rebellion, 'Maasai Evictions in Tanzania', XR Global Support: newsletter-no-reply@rebellion.earth; 'Giraffes, Parrots, and Oak Trees are among Many Species Facing Extinction', *WomMin Weekly Bulletin*, 25 August 2022.

27 Vandana Shiva (ed.), *Gates to the Global Empire*. Santa Fe, NM: Synergic Press, 2022.

28 'Take Action': https://abolish frontext.org.

29 Hannah Fair and Mathew McMullen, 'Toward a Theory of Non-Human Species-Being', *Environmental Humanities*, 15 (2023) 193–212.

30 Ariel Salleh (ed.), *Eco-Sufficiency and Global Justice: Women Write Political Ecology*. London: Pluto Press, 2009.

31 *Fight the Fire – Ecosocialist Magazine*: https://www.fightthefire.net.

32 Richard Peet, Paul Robbins, and Michael Watts (eds.), *Global Political Ecology*. London: Routledge, 2011; Raymond Bryant (ed.), *The International Handbook of Political Ecology*. Cheltenham, UK: Edward Elgar, 2015.

33 Peter Hay, *Main Currents in Western Environmental Thought*. Sydney: University of New South Wales Press, 2002.

34 Ashish Kothari, Ariel Salleh, Arturo Escobar, Federico Demaria, and Alberto Acosta (eds.), *Pluriverse: A Post-Development Dictionary*. New Delhi: Tulika and Authors UpFront, 2019.

35 Leah Temper, Daniela del Bene, Joan Martinez-Alier, 'Mapping the Frontiers and Front Lines of Global Environmental Justice: The EJAtlas', *Journal of Political Ecology* 22 (2015) 255–278: https://doi.org/10.2458/v22i1.21108; Ashish Kothari, 'Radical Ecological Democracy: A Path Forward for India and Beyond', *Development*, 57 (2014) 36–45; Geoffrey Pleyers, 'Alter-Globalization Movement'; Cormac Cullinan, 'Nature Rights' in Kothari et al. (eds.), 2019.

36 Leonardo Figueroa-Helland and Pratik Raghu, 'Indigeneity Vs. "Civilization": Indigenous Alternatives to The Planetary Rift' in J. Smith, M. Goodhart, P. Manning, and J. Markoff (eds.), *Social Movements and World System Transformation*. New York: Routledge, 2017.

37 Ariel Salleh, 'Global Alternatives and the Meta-Industrial Class' in R. Albritton, J. Bell, S. Bell, and R. Westra (eds.), *New Socialisms: Futures Beyond Globalization*. London: Routledge, 2004; Sylvia Tamale, *Decolonization and Afro-Feminism*. Carlton, CA: Daraja Press, 2020; Mumbi Maina-Okori, Jade Renee Koushik, and Alexandria Wilson, 'Reimagining Intersectionality in Environmental and Sustainability Education: A Critical Literature Review', *The Journal of Environmental Education*, 49 (2018) 266–96.

38 Greta Thunberg, 'There Are No Real Climate Leaders Yet – Who Will Step Up At COP26?', *The Guardian*, 21 October 2021: https://www.other-news.info/there-are-no-real-climate-leaders-yet-who-will-step-up-at-cop26/.

Chapter 2

Adapting an early text, later published as Ariel Salleh, 'Review Essay: Hossay's *Unsustainable* and Shiva's *Earth Democracy*', *Organization & Environment*, 19 (2006) 406–10.

1 Rudolf Bahro, *From Red to Green*. London: Verso, 1984; Wolfgang Sachs (ed.), *The Development Dictionary: A Guide to Knowledge as Power*. London: Zed Books, 1992/2010.

2 H. M. Vyas, *Village Swaraj*. Ahmedabad: Navajivan, 1962; Ashish Kothari and K. J. Joy (eds.), *Alternative Futures: India Unshackled*. New Delhi: AuthorsUpFront, 2017; Radical Ecological Democracy: www.radicalecologicaldemocracy.org.

3 Ivan Illich, *Energy and Equity*. New York: Boyars, 1972; Ivan Illich, *Tools for Conviviality*. New York: Harper and Row, 1993; Gustavo Esteva and Madhu Suri Prakash, *Grassroots Postmodernism: Remaking the Soil of Cultures*. London: Zed

Books, 1998; Manfred Max-Neef, Antonio Elizalde and Martin Hopenhyan, *Human Scale Development*. Development Dialogue, Uppsala: Cepaur-Dag Hammarskjöld Foundation, 1989; Pablo Solon, 'Notes for the Debate: Vivir Bien / Buen Vivir?', *Systemic Alternatives*, 2014: www.systemicalternatives.com.

4   Kirkpatrick Sale, 'Principles of Bioregionalism' in J. Mander and E. Goldsmith (eds.), *The Case Against the Global Economy*. San Francisco: Sierra Club, 1996.

5   Bill Mollison, *Permaculture: A Designers' Manual*. Tyalgum, NSW: Tagari Publications, 1988; Ted Trainer, *Abandon Affluence!* London: Zed Books, 1985; Terry Leahy, *The Politics of Permaculture*. London: Pluto, 2021.

6   Mogobe Ramose, *African Philosophy through Ubuntu*. Harare: Mond Books, 2002; Lesley Le Grange, 'Ubuntu' in A. Kothari, A. Salleh, A. Escobar, F. Demaria, and A. Acosta (eds.), *Pluriverse: A Post-Development Dictionary*. New Delhi: Tulika and AuthorsUpFront, 2019, p. 324; Inge Konik, 'Ubuntu and Ecofeminism', *Environmental Values*, 27 (2018) 269–88.

7   Kyle Whyte, 'From Planetary to Societal Boundaries: An Argument for Collectively Defined Self-Limitation', 23 July 2021: https://www.tandfonline.com/doi/full/10.1080/15487733.2021.1940754; Brian Tokar and Tamra Gilbertson (eds.), *Climate Justice and Community Renewal: Resistance and Grassroots Solutions*. London: Oxford University Press, 2020.

8   Sean McDonagh, *To Care for the Earth: A Call to a New Theology*. London: Chapman, 1986; Leonardo Boff, *Cry of the Earth, Cry of the Poor*. New York: Orbis Books, 1997; Pope Francis, *Laudato Si': On Care for Our Common Home*, 2015: http://w2.vatican.va/content/francesco/en/encyclicals/documents/papa-francesco_20150524_enciclica-laudato-si.html.

9   Dianne Rocheleau, Barbara Thomas-Slayter, and Esther Wangari (eds.), *Feminist Political Ecology: Global Issues and Local Experiences*. New York: Routledge, 1996.

10 Rosa Luxemburg, *The Accumulation of Capital*. New York: Monthly Review Press, 1913/1968; Maria Mies, *Patriarchy and Accumulation on a World Scale*. London: Zed Books, 1986; Ariel Salleh, *Ecofeminism as Politics: Nature, Marx, and the Postmodern*. London: Zed Books, 1997/2017; Mary Mellor, 'Ecofeminist Political Economy and the Politics of Money' in A. Salleh (ed.), *Eco-Sufficiency and Global Justice: Women Write Political Ecology*. London: Pluto Press, 2009.

11 Vandana Shiva, *Staying Alive: Women, Ecology, and Development*. London: Zed Books, 1989; Vandana Shiva and Kartikey Shiva, *Oneness vs the 1%: Shattering Illusions, Seeding Freedom*. North Geelong, VIC: Spinifex, 2018.

12 Vandana Shiva, *Earth Democracy: Justice, Sustainability, and Peace*. Cambridge, MA: South End, 2006, p. 161.

13 Shiva, 2006, p. 6.

14 Shiva, 2006, p. 3.

15 Patrick Hossay, *Unsustainable: A Primer for Global Environmental and Social Justice*. London: Zed Books, 2006.

16 American Association for the Advancement of Science, *The FAO Measure of Chronic Undernourishment: What is it Really Measuring?* Washington, DC: AAAS, 1997, p. 6.

17 Vandana Shiva, *Water Wars*. Cambridge, MA: South End, 2003, p. 33.

18 Melinda Hinkson, *See How We Roll: Enduring Exile between Desert and Urban Australia*. Durham, NC: Duke University Press, 2021.

19 Shiva and Shiva, 2018.

Chapter 3

Adapting an early text, later published as '*Terra Nullius*' in *Ecofeminism as Politics: Nature, Marx, and the Postmodern*. London: Zed Books, 1997/2017.

1 Stephan Schmidheiny (ed.), *Changing Course: A Global Business Perspective on Development and the Environment*. Cambridge, MA: MIT Press, 1992; Bob Burton, 'Right Wing Think Tanks go Environmental', *Chain Reaction*, 73 (1995); Kenny Bruno, *The Greenpeace Book of Greenwash*. Washington: HEIP Campaign, 1992; Ariel Salleh, 'Some Reflections on Our Political Times', *The Ecofeminist Newsletter*, No. 4, September 1993.

2 Ariel Salleh, 'Politics in/of the Wilderness', *Arena Magazine*, 23 (1996) 26–30; Christine Christopherson with Marcia Langton, 'Allarda! No to the Ranger Uranium Mine', *Arena Magazine*, 17 (1995) 28–32; Richard Ledgar, 'Links between Ranger Uranium Mining and France's Nuclear Program', *Newsletter of the Environment Centre of the Northern Territory*, August 1995.

3 Workshop on Regional Agreements, EcoPolitics IX Conference, Darwin, 1998.

4 Friends of the Earth (FOE), 'Special Issue: Black Lives Matter', *Chain Reaction*, 141 (2021).

5 Helen Rosenbaum (ed.), *Principles for the Environmental Management of Australian Mining Companies Operating in Papua New Guinea*. Melbourne: ACF, 1995; and WEEP, PO Box 4830 Boroko, NCD, Papua New Guinea.

6 Malini Karkal, *Can Family Planning Solve the Population Problem?* Bombay: Stree Uvach, 1989; Pat Hynes, *Taking Population Out of the Equation*. Amherst: Institute on Women and Technology, 1993; Betsy Hartmann, *Reproductive Rights and Wrongs: The Global Politics of Population Control and Reproductive Choice*. New York: Harper, 1987.

7. Lynette Dumble, 'Women and the UN: Another Forged Consensus?', *Green Left Weekly*, 20 September 1995, p. 3; 'Population Control's Medical Paradigm: Regulation of Fertility or Disruption of Lives', *Newsletter: Women's Global Network for Reproductive Rights*, 50 (1995) pp. ii–iv.

8. Lynette Dumble, 'Population Control or Empowerment of Women?', *Green Left Weekly*, 2 November 1994, p. 15; John Clark, *The Tragedy of Common Sense*. Regina, SK: Changing Suns Press, 2016.

9. Quentin Beresford, *Working Country*. Sydney: New South Press, 2021.

10. George Caffentzis. 'The Fundamental Implications of the Debt Crisis for Social Reproduction in Africa' in M. and G. Dalla Costa (eds.), *Paying the Price*. London: Zed Books, 1995, p. 19; italics added.

11. Caffentzis, 1995, p. 31.

12. Silvia Federici, 'The Debt Crisis, Africa and the New Enclosures', in Midnight Notes Collective (eds.), *Midnight Oil: Work, Energy, War*. New York: Autonomedia, 1992.

13. Michael Chossudovsky, 'IMF-World Bank Policies and the Rwandan Holocaust', *Third World Resurgence*, December 1994; Sam Kiley, 'UK Firm in Rwanda Arms Trade', *The Australian*, 11 November 1996.

14. Henrietta Fourmile, 'Protecting Indigenous Intellectual Property Rights in Biodiversity', *EcoPolitics IX Conference Papers and Resolutions*, Darwin 1995.

15. Editor, 'Biopiracy Update', *Pacific News Bulletin*, January 1996.

16. Darrell Posey, 'Indigenous Peoples and Traditional Resource Rights: A Basis for Equitable Relationships?', *EcoPolitics IX Conference Papers and Resolutions*, 1998.

17. Michael Dodson, 'Indigenous Peoples and Intellectual Property Rights', EcoPolitics IX Conference Papers and Resolutions, 1998.

18. Bill Freeland, 'Workshop on Intellectual Property Rights', EcoPolitics IX Conference Papers and Resolutions, 1998.

19. 'Second Conference of the Parties to the Convention on Biological Diversity', Jakarta, 6–17 November 1995; Bob Phelps (ed.), *Newsletter of the Australian GeneEthics Network*: info@geneethics.org.

20. Marshall Sahlins, *Stone Age Economics*. New York: Aldine, 1972.

21. Bruce Pascoe, *Dark Emu: Aboriginal Australia and the Birth of Agriculture*. Broome, WA: Magabala Books, 2014, p. 30.

22. Jon Altman, 'The Native Title Act Supports Mineral Extraction and Heritage Destruction', *Arena Online*, 16 June 2020: https://arena.org.au/the-native-title-act-supports-mineral-extraction-and-heritage-destruction/.

23. Dan Tout, 'Juukan Gorge Destruction', *Arena*, 4 (2020) 61–7, p. 61.

24 Kirsty Howey, 'Gas Town, Darwin', *Arena*, 15 (2023) 53–9.

25 First Nations National Constitutional Convention, 'Uluru Statement from the Heart': https://ulurustatement.org.

26 Pip Hinman and Ruth Heyman, 'Celeste Liddell on the Voice: Truth-telling is the First Important Step', *Green Left*, 26 September 2023.

27 Ashish Kothari, Ariel Salleh, Arturo Escobar, Federico Demaria, and Alberto Acosta (eds.), *Pluriverse: A Post-Development Dictionary*. New Delhi: Tulika and AuthorsUpFront, 2019; Tyson Yunkaporta, *Sand Talk*. Melbourne: Text Publishing, 2019.

Chapter 4

Adapting Ariel Salleh, 'Fukushima: A Call for Women's Leadership', *The New Significance*, 1 November 2011.

1 Jim Green, 'Fukushima: The Political Fallout in Australia', *Chain Reaction Magazine*, 112 (2011).

2 Roger Pulvers, 'Broadcast Transcript: 'Japan After its Triple Disaster of 2011', *The Science Show*, Australian Broadcasting Corporation, ABC Radio National, 23 July 2011.

3 Michael Renner, 'Assessing the Military's War on the Environment' in L. Brown (ed.), *State of the World Report*. New York: Norton, 1991.

4 Helen Caldicott, 'Unsafe at Any Dose', *New York Times*, 30 April 2011: http://www.nytimes.com/2011/05/01/opinion/01caldicott.html.

5 Peter Karamoskos and Jim Green, 'Do We Know the Chernobyl Death Toll?', *Chain Reaction Magazine*, 112 (2011), p. 23.

6 Peter Karamoskos, 'Radiating Risk and Undermining Public Health', *Online Opinion*, 13 December 2010: http://www.onlineopinion.com.au/view.asp?article=11358.

7 Whitney Graham and Elena Nicklasson, 'Maternal Meltdown from Chernobyl to Fukushima', Global Movement for Children, San Francisco, 26 April 2011.

8 Mariko Sanchanta and Mitsuri Obe, 'Moms Turn Activists in Japanese Crisis', *Wall Street Journal*, 17 June 2011.

9 Pulvers, ABC Broadcast, 2011.

10 Chigaya Kinoshita, 'The Shock Doctrine of Japanese Type: Neoliberalism and the Shadow of America', 29 May 2011: http://www.jfissures.org/.

11 Keitaro Morita, 'For a Better Environmental Communication: A Materialist Ecofeminist Analysis of Global Warming', Rikkyo University, Tokyo, 2006: www.eca.usp.br/caligrama/english/06_keitaro.pdf.

12　Dorothy Nelkin, 'Nuclear Power as a Feminist Issue', *Environment,* 23 (1981) 14–24; Mary Goebel Noguchi, 'The Rise of the Housewife Activist', *Yomiuri Shimbun,* September 1992.

13　Mike Danaher, 'On the Forest Fringes? Environmentalism, Left Politics and Feminism in Japan', *Transformations,* 2003: http://transformations.cqu.edu.au/journal/issue_06/pdf/danaher.pdf.

14　Friends of the Earth, *Chain Reaction Magazine,* 3 (1978).

15　Susan Griffin, *Woman and Nature: The Roaring Inside Her.* New York: Harper, 1978; Elizabeth Dodson Gray, *Green Paradise Lost.* Wellesley, MA: Roundtable Press, 1979.

16　Joyce Cheney, 'The Boys Got Us into This Mess', *Commonwoman,* 1979, quoted in Nelkin, 1981, p. 38.

17　Manushi Collective, 'Drought: God Sent or Man-Made Disaster?', *Manushi Newsletter,* No. 6, 1980.

18　Carolyn Merchant, *The Death of Nature: Women, Ecology and the Scientific Revolution.* San Francisco: Harper, 1980.

19　Lynne Jones (ed.), *Keeping the Peace.* London: Women's Press, 1983; Alice Cook and Gwyn Kirk, *Greenham Women Everywhere.* London: Pluto, 1983.

20　Leonie Caldecott and Stephanie Leland (eds.), *Reclaim the Earth.* London: Women's Press, 1983.

21　Ariel Salleh, 'Deeper than Deep Ecology: The Ecofeminist Connection', *Environmental Ethics,* 6 (1984) 335–41.

22　Petra Kelly, *Fighting for Hope.* London: Chatto and Windus, 1984.

23　Maria Mies, *Patriarchy and Accumulation on a World Scale.* London: Zed Books, 1986; Chellis Glendinning, *Waking Up in the Nuclear Age.* New York: Morrow, 1987.

24　Women Working for a Nuclear Free and Independent Pacific (eds.), *Pacific Women Speak.* Oxford: Greenline, 1987.

25　Carolyn Merchant, *Earthcare.* New York: Routledge, 1995.

26　Vandana Shiva, *Staying Alive: Women, Ecology and Development.* London: Zed Books, 1989.

27　Chris Cuomo, 'Still Fooling with Mother-Nature', *Hypatia,* 16 (2001) 149–55; Sherilyn MacGregor, 'From Care to Citizenship: Calling Ecofeminism Back to Politics', *Ethics and the Environment,* 9 (2004) 56–84.

28　Stephan Schmidheiny (ed.), *Changing Course: A Global Business Perspective on Development and the Environment.* Cambridge, MA: MIT Press, 1992.

29 Jed Greer and Kenny Bruno, *Greenwash: The Reality Behind Corporate Environmentalism*. Penang: Third World Network, 1996.

30 Gerry Canavan, Lisa Klarr, and Ryan Vu, 'Embodied Materialism in Action: Interview with Ariel Salleh', *Polygraph*, 22 (2010) 183–99.

31 'First Continental Summit of Indigenous Women', *Lucha Indigena,* Llapa Runaq Hatariynin, 34-Inti Raymi, 2009: translation by Marilyn Obeid, Sydney.

32 Ariel Salleh, 'Climate Strategy: Making the Choice between Ecological Modernisation or Living Well', *Journal of Australian Political Economy*, 66 (2011) 124–49.

33 Cormac Cullinan, 'The Universal Declaration of the Rights of Mother Earth' in M. Barlow (ed.), *Does Nature Have Rights?* Ottawa: Council of Canadians, 2010.

34 Silvia Federici and George Caffentzis, 'Must We Rebuild Their Anthill?': http://jfissures.wordpress.com/2011/04/22.

35 Women and Life-on-Earth, 'Obituary for Satomi Oba': http://www.wloe.org/Remembering-Satomi-Oba.513.0.html.

36 Michael Chandler, 'In Japan, New Attention for Longtime Anti-Nuclear Activist', *Washington Post*, 11 April 2011.

37 Gender and Climate Change: www.GenderCC.org; Meike Spitzner, 'How Global Warming is Gendered' in Ariel Salleh (ed.), *Eco-Sufficiency and Global Justice*: *Women Write Political Ecology*. London: Pluto Press, 2009.

38 NOW, Media Release: 'Spike in Infant Mortality in the Northwest Linked to Radiation Fallout from Fukushima Nuclear Power Plant Disaster', 16 June 2011: www.canow.org.

39 Asia Pacific Forum on Women, Law and Development, Letter to Prime Minister Mr. Naoto Kan, Prime Minister of Japan, 7 July 2011: www.apwld.org/.

40 Julian Ryall, 'Japan's Plan to Release Water from Fukushima Angers Fishermen', *Sydney Morning Herald*, 17 October 2020.

41 Stockholm International Peace Research Institute (SIPRI): https://www.sipri.org/sites/default/Bles/2020-06/yb20_summary_en_v2.pdf.

42 International Campaign to Abolish Nuclear Weapons: https://www.icanw.org/entryintoforce.

43 The Australian Naval Nuclear Power Safety (Transitional Provisions) Bill, 2023.

44 Simon Butler, '10 Reasons Why Climate Activists should not Support Nuclear', *Green Left Weekly,* June 2021, p. 12; Nic Maclellan, 'AUKUS Disrupts a Very Peaceful Part of Planet Earth', *Chain Reaction Magazine,* 141 (2021).

Chapter 5

A later version of this text is published in Raymond Bryant (ed.), 'Neoliberalism, Scientism, and Earth System Governance', *International Handbook of Political Ecology*. Cheltenham: Edward Elgar, 2015.

1. Manuel Castells, *Rise of the Network Society*. New York: Wiley, 2000; Michael Hardt and Antonio Negri, *Multitude*. New York: Penguin, 2004.

2. Bruno Latour, *Reassembling the Social*. Oxford: Oxford University Press, 2005.

3. Donella Meadows, Dennis Meadows, Jorgen Randers, and William Behrens, *The Limits to Growth*. New York: Universe, 1972.

4. Ulrich Brand and Markus Wissen, 'Global Environmental Politics and the Imperial Mode of Living', *Globalizations*, 9 (2012) 547–60; Patrick Bond, 'Post-imperialist North-South Financial Relations?' *Studies in Political Economy*, 81 (2008) 77–97.

5. Ariel Salleh, 'Ecosocialism, Gendered Imaginaries, and the Informatic-Securitization Complex', *Capitalism Nature Socialism*, 25 (2014) 24–39.

6. Maristella Svampa, 'Resource Extractivism and Alternatives', *Journal für Entwicklungspolitik*, 28 (2012) 43–73.

7. James Goodman and Ariel Salleh, 'The Green Economy', *Globalizations*, 10 (2013) 343–56.

8. Markus Kroger, *Extractivisms, Existences, and Extinctions*. London: Routledge, 2022.

9. Paul Robbins, *Political Ecology*. Oxford: Wiley-Blackwell, 2012.

10. Frank Biermann and Philipp Pattberg (eds.), *Global Environmental Governance Reconsidered*. Cambridge, MA: MIT Press, 2012, p. 265.

11. Gerry Canavan, Lisa Klarr, and Ryan Vu, 'Embodied Materialism in Action: Interview with Ariel Salleh', *Polygraph*, 22 (2010) 183–99.

12. George Kennan, 'To Prevent a World Wasteland', *Foreign Affairs*, 48 (1970) 401–13.

13. World Commission on Environment and Development (WCED), *Our Common Future*. Oxford: Oxford University Press, 1987.

14. Stephan Schmidheiny (ed.), *Changing Course: A Global Business Perspective on Development and the Environment*. Cambridge, MA: MIT Press, 1992.

15. Rosa Luxemburg, *The Accumulation of Capital*. New York: Monthly Review Press, 1913/1968.

16. Tebtebba Foundation, 'Indigenous Peoples' Seattle Declaration on the Third Ministerial Meeting of the World Trade Organization', 8 December 1999.

17 Frank Biermann, *Needed Now*. Berlin: Wissenschaftszentrum Berlin für Sozialforschung GmbH, 1998.

18 Biermann and Pattberg (eds.), 2012, p. xii.

19 William Carroll, 'Hegemony and Counter-Hegemony in a Global Field', *Studies in Social Justice*, 1 (2007) 36–66; William Carroll and Jean-Paul Sapinski, 'Neoliberalism and the Transnational Capitalist Class' in S. Springer, K. Birch, and J. MacLeavy (eds.), *The Handbook of Neoliberalism*. New York, NY: Routledge, 2016.

20 Oran Young, 'The Architecture of Global Environmental Governance', *Global Environmental Politics*, 8 (2008) 14–32, p. 21.

21 Daniel Bromley, 'Environmental Governance as Stochastic Belief Updating', *Ecology and Society*, 17 (2012) 14, pp. 1–2.

22 Leslie Sklair, *The Transnational Capitalist Class*. Oxford: Blackwell, 2001; K. Tienhaara, A. Orsini, and R. Falkner, 'Global Corporations' in Biermann and Pattberg (eds.), 2012.

23 Biermann and Pattberg (eds.), 2012, p. 274.

24 Frank Biermann, 'The Rationale for a World Environment Organization' in F. Biermann and S. Bauer (eds.), *Global Environmental Governance*, Burlington, VT: Ashgate, 2005.

25 Max Horkheimer and Theodor Adorno, *Dialectic of Enlightenment*. London: Penguin, 1973.

26 Frank Biermann, 'Planetary Boundaries and Earth System Governance', *Ecological Economics*, 81 (2012) 4–9, p. 4.

27 Evelyn Fox Keller, *Reflections on Gender and Science*. New Haven, CT: Yale University Press, 1985; Vandana Shiva, *Staying Alive: Women, Ecology, and Development*. London: Zed Books, 1989.

28 Robert May, S. Levin, and G. Sugihara, 'Complex Systems: Ecology for Bankers', *Nature*, 451 (2008) 893–5, p. 893.

29 Bromley, 2012, p. 2; italics added.

30 Brian Wynne, 'Methodology and Institutions of Value as Seen from the Risk Field' in J. Foster (ed.), *Valuing Nature?* London: Routledge, 1997.

31 Ulrich Beck, Anthony Giddens, and Scott Lash, *Reflexive Modernization*. Stanford, CA: Stanford University Press, 1994.

32 Arthur Mol, David Sonnenfeld, and Gert Spaargaren (eds.), *The Ecological Modernisation Reader*. London: Routledge, 2009.

33 Paul Hawken, Amory Lovins, and Hunter Lovins, *Natural Capitalism*. London: Earthscan, 1999.

34 Richard York and Eugene Rosa, 'Key Challenges to Ecological Modernization Theory', *Organization & Environment*, 16 (2003) 273–88; John Bellamy Foster, 'The Planetary

Rift and the New Human Exemptionalism', *Organization & Environment*, 25 (2012) 211–37.

35 Ariel Salleh, 'Organised Irresponsibility', *Environmental Politics*, 15 (2006) 388–416.

36 Ted Trainer, *Abandon Affluence!* London: Zed Books, 1985.

37 Johann Rockström, W. Steffen, K. Noone, A. Persson, F. Chapin, E. Lambin, T. Lenton, M. Scheffer, C. Folke, H. Schellnhuber, B. Nykvist, C. de Wit, T. Hughes, S. van der Leeuw, H. Rodhe, S. Sörlin, P. Snyder, R. Costanza, U. Svedin, M. Falkenmark, L. Karlberg, R. Corell, V. Fabry, J. Hansen, B. Walker, D. Liverman, K. Richardson, P. Crutzen, and J. Foley, 'A Safe Operating Space for Humanity', *Nature*, 461 (2009) 472–75.

38 German Advisory Council on Global Change, *World in Transition*. London: Earthscan, 2012, p. 3.

39 Biermann, 2012, p. 5.

40 Biermann, 2012, p. 6.

41 Biermann, 2012, p. 7.

42 Biermann, 2012, p. 6.

43 V. Galaz, F. Biermann, C. Folke, M. Nilsson, and P. Olsson, 'Global Environmental Governance and Planetary Boundaries', *Ecological Economics*, 81 (2012) 1–3.

44 M. Padmanabhan and S. Jungcurt, 'Bio-complexity: Conceptual Challenges for Institutional Analysis in Biodiversity Governance', *Ecological Economics*, 81 (2012) 70–9, p. 70; italics added.

45 Padmanabhan and Jungcurt, 2012, p. 73.

46 Padmanabhan and Jungcurt, 2012, p. 72.

47 Padmanabhan and Jungkurt, 2012, p. 74 citing Anil Agrawal, 'Common Property Institutions and Sustainable Governance of Resources', *World Development*, 29 (2001) 1649–72.

48 Elinor Ostrom, 'Scales, Polycentricity and Incentives' in L. Guruswamy and J. McNeely (eds.), *Protection of Global Biodiversity*. Durham, NC: Duke University Press, 1998.

49 Padmanabhan and Jungkurt, 2012, p. 75 quoting Kurt Hagerdorn, 'Particular Requirements for Institutional Analysis in Nature-Related Sectors', Keynote, XII Congress of the European Association of Agricultural Economists, Ghent, Belgium, August 2008, p. 12.

50 Galaz et al., 2012, p. 1.

51　Ariel Salleh, Mary Mellor, Kathryn Farrell, and Vandana Shiva, 'How Ecofeminists Use Complexity in Ecological Economics' in K. Farrell, T. Luzzati, and S. van den Hove (eds.), *Beyond Reductionism*. London: Routledge, 2013.

52　M. Nilsson and A. Persson, 'Can Earth Systems be Governed?' *Environmental Economics*, 81 (2012) 10–20, p. 18; italics added.

53　L. Partzsch and R. Ziegler, 'Social Entrepreneurs as Change Agents', *International Environmental Agreements*, 11 (2011) 63–83.

54　Timothy Luke, 'Sustainable Development as a Power/Knowledge System' in F. Fisher and M. Black (eds.), *Greening Environmental Policy*. New York: St Martins, 1995, p. 30; Michel Foucault, *Power/Knowledge*. New York: Pantheon, 1980.

55　Marilyn Waring, 'Policy and the Measure of Woman: UNSNA, ISEW, HDI, and GPI'; and Mary Mellor, 'Ecofeminist Political Economy and the Politics of Money' in Ariel Salleh (ed.), *Eco-Sufficiency and Global Justice: Women Write Political Ecology*. London: Pluto Press, 2009.

56　Mujeres Manifesto, 'First Continental Summit of Indigenous Women', *Lucha Indígena*, 34, 2009: www.luchaindigena.com/; Ralph Regenvanu, 'The Traditional Economy as Source of Resilience in Vanuatu' in T. Anderson and G. Lee (eds.), *In Defense of Melanesian Customary Land*. Sydney: AidWatch, 2010.

57　Andrew Stirling, 'Multi-Criteria Mapping' in J. Foster (ed.), *Valuing Nature?* London: Routledge, 1997.

58　Kees Van der Pijl, *Transnational Classes and International Relations*. London: Routledge, 1998.

59　Aarti Gupta, Steinar Andreson, Bernd Siebenhuner, and Frank Biermann, 'Science Networks' in Biermann and Pattberg (eds.), 2012.

60　Silvio Funtowicz and Jerry Ravetz, 'Science for the Post-Normal Age', *Futures*, 25 (1993) 739–55.

61　Sheila Jasanoff, 'Testing Time for Climate Science', *Science*, 328 (2010) 695–6; Ariel Salleh, 'Editorial: A Sociological Reflection on the Complexities of Climate Change Research', *International Journal of Water*, 5 (2010) 285–97.

62　John Bellamy Foster, Brett Clark, and Richard York, *The Ecological Rift*. New York: Monthly Review, 2010.

63　Luke, 1995, p. 30.

64　Manfred Max-Neef, Antonio Elizalde, and Martin Hopenhyan, *Human Scale Development. Development Dialogue*, Uppsala: Cepaur-Dag Hammarskjöld Foundation, 1989; Veronika Bennholdt-Thomsen and Maria Mies, *The Subsistence Perspective*. London: Zed Books, 1999; Ariel Salleh, 'From Metabolic Rift to Metabolic Value: Reflections on Environmental Sociology and the Alternative Globalization Movement', *Organization & Environment*, 23 (2010) 205–19.

Chapter 6

Adapting 'Risky Science: Can we make our *Gene Technology Act* Ethical?' *New Matilda*, 1 June 2005; and 'A Splice of Life: How Do We Regulate Novel Organisms?' *New Matilda*, 15 June 2005: www.newMatilda.com.

1. Erwin Schrodinger, *What is Life? The Physical Aspects of the Living Cell*. New York: Macmillan, 1945; Craig Venter, 'What is Life?', Address to Euroscience Open Forum, Dublin, 14 July 2012.

2. Stuart Newman, 'The Demise of the Gene', *Capitalism Nature Socialism*, 24 (2013) 62–72, p. 65.

3. Tina Stevens and Stuart Newman, *Biotech Juggernaut: Hope, Hype, and Hidden Agendas of Entrepreneurial Biotech*. London: Routledge, 2020.

4. Office of the Gene Technology Regulator (OGTR), *Handbook on the Regulation of Gene Technology in Australia*. Canberra: AGPS, 2001: www.ogtr.gov.au; Ariel Salleh, 'Organised Irresponsibility: Contradictions in the Australian Government's Strategy for GM Regulation', *Environmental Politics*, 15 (2006) 399–416.

5. A. Lazaris, S. Arcidiacono, Y. Huang, J. Zhou, F. Duguay, and N. Chretien, 'Spider Silk Fibers Spun from Soluble Recombinant Silk Produced in Mammalian Cells', *Science*, 295 (2002) 472–6.

6. Commonwealth Scientific and Industrial Research Organization (CSIRO), 'Inductive Hazard Analysis for GMOs'. Canberra: April 2003: www.biodiversity.csiro.au/2nd_level/3rd_level.plan_gmos.htm.

7. Les Levidow and Susan Carr, 'Unsound Science? Trans-Atlantic Regulatory Disputes Over GM Crops', *International Journal of Biotechnology*, 2 (2000) 257–73.

8. James Watson and Francis Crick, 'A Structure for Deoxyribose Nucleic Acid', *Nature*, 171 (1953) 737–8.

9. Brian Goodwin, 'A Relational or Field Theory of Reproduction' in M. Wan Ho and P. Saunders (eds.), *Beyond NeoDarwinism*. London: Academic Press, 1984; Ruth Hubbard and George Wald, *Exploding the Gene Myth*. Boston: Beacon, 1993; Vandana Shiva and Ingunn Moser (eds.), *Biopolitics*. London: Zed Books, 1995; Barry Commoner, 'Unravelling the DNA Myth', *Harpers*, February 2002.

10. Sheila Jasanoff, 'Bridging the Two Cultures of Risk Analysis', *Risk Analysis*, 113 (1993) 123–9.

11. Dire Tladi, 'The Biosafety Protocol and the Promotion of Sustainable Development: With One Hand it Giveth, with the Savings Clause it Taketh Away?' *The Comparative and International Law Journal of Southern Africa*, 39 (2006) 83–101; Lim Li Lin and Lim Li Ching, 'The Cartagena Biosafety Protocol is Just the Beginning', *Third World Resurgence*, 161 (2004) 26–31; Anna Salleh, 'Applying the

Precautionary Principle', *ABC Science Online*, 20 December 2001: www.abc.net.au/science.

12 'Biological Complexity': www.phy.auckland.ac.nz/staff/prw/biocomplexity/summary.htm.

13 Silvio Funtowicz and Jerome Ravetz, 'Post-Normal Science: Environmental Policy Under Conditions of Complexity', 2004: www.nusap.net; Andrea Saltelli and Monica Di Fiore (eds.), *The Politics of Modelling: Numbers Between Science and Policy*. Oxford: OUP Academic, 2023.

14 Erosion, Technology, Concentration (ETC), 'Syngenta Claims Multi-Genome Monopoly', *ETC News Release,* 10 January 2005.

15 GeneEthics, 'Media Release', GE Canola Should Be Banned', 25 July 2003: www.geneethics.org.

16 *Gene Technology Act 2000, No. 169* and *Gene Technology Regulations 2001, Statutory Rules No. 106*. Canberra: AGPS, 2001.

17 'Kari-Oca Declaration and Indigenous Peoples Earth Charter' in L. Van der Vlist (ed.), *Voices of the Earth*. Amsterdam: International Books, 1994.

18 Shiv Visvanathan, 'Knowledge, Justice and Democracy' in M. Leach, I. Scoones, and B. Wynne (eds.), *Science and Citizens*. London: Zed Books, 2005.

19 Britta van Beers, 'Rewriting the Human Genome, Rewriting Human Rights Law?', *Journal of Law and the Biosciences*, 7 (2020), 1–36: https://doi.org/10.1093/jlb/Isaa006.

20 Melinda Cooper, *Life as Surplus: Biotechnology and Capitalism in the Neoliberal Era*. Seattle: University of Washington Press, 2008; George Church and Ed Regis, *Regenesis: How Synthetic Biology Will Reinvent Nature and Ourselves*. New York: Basic Books, 2012.

21 Renate Klein, 'Reproductive Engineering' in A. Kothari, A. Salleh, A. Escobar, F. Demaria, and A. Acosta (eds.), *Pluriverse: A Post-Development Dictionary*. New Delhi: Tulika and AuthorsUpFront, 2019, p. 67.

22 Alliance for Human Research Protection, 'Advancing Voluntary, Informed Consent to Medical Intervention: What is Gain of Function Research? Who is at High Risk?', 19 May 2020: https://ah rp.org/.

23 USA Today, 'Inside America's Secretive Biolabs', 28 May 2015: https://www.usatoday.com/story/news/2015/05/28/biolaba-pathogens-location-incidents/26587505/.

24 GAVI, 'World Leaders Commit to GAVI's Vision to Protect the Next Generation with Vaccines', 23 January 2020: https://www.gavi.org/news/media-room/world-leaders-commit-gavis-vision-protect-next-generation-vaccines.

25 GMWatch, 20 May 2021: http://www.gmwatch.org/en/news/latest-news/19403-wuhan-and-us-scientists-used-undetectable-methods-of-genetic engineering-on-bat-corona-viruses.

26 M-CAM: https://www.m-cam.com; David Martin, 'The Fauci/Covid-19 Dossier'. CC-BY-NC-SA, 2021, p. 16.

27 Megan Redshaw, 'CDC "Corrects" Number of Reported Deaths After COVID Vaccines by Dumping Foreign Reports', *Global Research*, 23 July 2021: https://www.globalresearch.ca/cdc-corrects-number-of-reported-deaths-after-covid-vaccines-by-dumping-foreign-reports/5750948.

28 Filippa Lentzos and Guy Reeves, 'Scientists are Working on Vaccines that Spread like a Disease', *Bulletin of Atomic Scientists*, 18 September 2020: https://thebulletin.org/2020/09/scientists-are-working-on-vaccines-that-spread-like-a-disease/.

29 Nuffield Council on BioEthics, 'Genome Editing and Human Reproduction: Social and Ethical Issues', No. III, 2018, p. 47.

Chapter 7

Adapting an early version of text, later published as Ariel Salleh, 'Climate Strategy: Making the Choice between Ecological Modernisation or "Living Well"', *Journal of Australian Political Economy*, 66 (2010) 124–49.

1 Bundesministerium für Umwelt, 'Naturschutz und Reaktorsicherheit, Ökologische Industriepolitik: Memorandum für einen "New Deal" von Wirtschaft, Umwelt und Beschäftigung, 2006, cited in Judith Delheim, 'Seven Theses for a Discussion on Energy Policy and Social-Ecological Conversion', unpubl ms. Berlin, 2007, pp. 9–11.

2 Martin Hajer, 'Ecological Modernisation as Cultural Politics' in S. Lash, B. Szerszinski, and B. Wynne (eds.), *Risk, Environment, and Modernity*. London: Sage, 1996.

3 Arthur Mol and David Sonnenfeld (eds.), *Ecological Modernisation Around the World*. London: Frank Cass, 2000; David Sonnenfeld (ed.), *The Ecological Modernization Reader: Environmental Reform in Theory and Practice*. New York: Routledge, 2009.

4 Richard York and Eugene Rosa, 'Key Challenges to Ecological Modernization Theory', *Organization & Environment*, 16 (2003) 273–88.

5 Gro Harlem Brundtland (ed.), *Our Common Future*. Geneva: WCED, 1987.

6 Sharon Astyk, 'A New Deal or a War Footing? Ruminations for a New Future', *Casuabon's Book*, 11 November 2008.

7 Via Campesina, 'Thousands of Cancuns for Climate Justice!', 2010: viacampesina@viacampesina.org.

8 Stuart Rosewarne and James Goodman, 'Beyond the CPRS', *The Age: National Times*, 9 December 2009.

9. Ariel Salleh (ed.), *Eco-Sufficiency and Global Justice: Women Write Political Ecology*. London: Pluto Press, 2009.

10. International Trade Union Confederation (ITUC), 2010: www.ituc-csi/international-trade-unions-to.html.

11. John Bellamy Foster, *Marx's Ecology*. New York: Monthly Review, 2000.

12. Gerd Johnsson-Latham, *Initial Study of Lifestyles, Consumption Patterns, Sustainable Development and Gender*. Stockholm: Swedish Ministry of Sustainable Development, 2006; Ana Isla, 'Who Pays for Kyoto Protocol?' in Salleh (ed.), 2009.

13. Climate Connections, 'Shell Bankrolls REDD', 2010: aa@globaljusticeecology.org.

14. International Trade Union Confederation (ITUC), 2010: www.ituc-csi/international-trade-unions-to.html.

15. Richard Heinberg, *Searching for a Miracle: Net Energy Limits and the Fate of Industrial Society*. San Francisco: International Forum on Globalization and the Post Carbon Institute, 2009.

16. Beyond Zero Emissions, *Zero Carbon Australia: Stationary Energy Plan*. Australia: Energy Research Institute, University of Melbourne: beyondzeroemissions.org.

17. Leigh Ewbank, 'The Transition Decade: Reshaping Australia's Climate Politics', *Chain Reaction Magazine*, 109 (2010).

18. IBON, *Primer on the Climate Crisis: Roots and Solutions*. Quezon City: IBON International, 2010.

19. Rising Tide, 2010: www.risingtidenorthamerica.org.

20. Petra Hesslerová and Jan Pokorny, 'Forest Clearing, Water Loss, and Land Surface Heating, as Development Costs', *International Journal of Water*, 5 (2010) 401–18.

21. Heather Gautney, *Protest and Organization in the Alternative Globalization Era*. New York: Palgrave, 2010; Ariel Salleh, 'Living with Nature: Reciprocity or Control?' in R. Engel and J. Engel (eds.), *Ethics of Environment and Development*. London: Pinter 1990; Ariel Salleh, 'Interview with Maria Mies: Women, Nature, and the International Division of Labour', *Science as Culture*, 9 (1990) 73–87; Veronika Bennholdt-Thomsen and Maria Mies, *The Subsistence Perspective*. London: Zed Books, 1999.

22. Evo Morales, 'Peoples' World Conference on Climate Change and Mother Earth Rights', Cochabamba, 19–22 April 2010: www.boliviaun.org/cms.

23. CMPCC Working Group 13, 'Intercultural Dialogue to Share Knowledge, Skills and Technologies', World People's Conference on Climate Change and the Rights of Mother Earth, Cochabamba, April 2010: boletin@cmpcc.org.

24. CMPCC Working Group 13, 2010, Clause 41.

25. CMPCC Working Group 13, 2010, Clause 42.

26 CMPCC Working Group 13, 2010, Clause 43.

27 IBON People's Protocol, *IBON Primer on Climate Change*. Quezon City: IBON International 2008.

28 CMPCC Working Group 13, 2010, Clause 45.

29 CMPCC Working Group 13, 2010, Clause 46.

30 CMPCC Working Group 13, 2010, Clause 50.

31 Ariel Salleh, 'Climate Change and the Other Footprint', *The Commoner*, 13 (2008) 103–13.

32 Third World Network (TWN), 'Divergent Views on Bodies of the UNFCCC Technology Mechanism', 2010: www.twnside.org.sg.

33 UNFCCC, 'Ad Hoc Working Group on Long-Term Cooperative Action under the Convention Twelfth Session Tianjin', October 2010, p. 43: http://unfccc.int/resource/docs/2010/awglca12/eng/14.pdf.

34 Global Environment Facility (GEF), 'Implementation of the Poznan Strategic Programme on Technology Transfer', 2010: www.thegef.com/gef/ccpublist.

35 CMPCC, 'Submission by the Plurinational State of Bolivia To the Ad Hoc Working Group on Long-Term Cooperative Action', 2010: boletin@cmpcc.org.

36 Ralph Regenvanu, 'The Traditional Economy as Source of Resilience in Vanuatu' in T. Anderson and G. Lee (eds.), *In Defence of Melanesian Customary Land*. Sydney: AidWatch, 2010.

37 People and Water NGO, *Košice Civic Protocol on Water, Vegetation and Climate Change and COP15*. Kosice: People and Water NGO, 2010, p. 1.

38 Wilhelm Ripl, 'Losing Fertile Matter to the Sea: How Landscape Entropy Affects Climate', *International Journal of Water*, 5 (2010) 353–64; Duane Norris and Peter Andrews, 'Recoupling the Carbon and Water Cycles by Natural Sequence Farming', *International Journal of Water*, 5 (2010) 386–95.

39 Fred Pearce, 'Rainforests May Pump Winds Worldwide', *New Scientist*, 2702 (2009).

40 Ariel Salleh, 'From Metabolic Risk to Metabolic Value: Reflections on Environmental Sociology and the Alternative Globalization Movement', *Organization & Environment*, 23 (2010) 205–19; Annie James and Neena Pathak Broome, 'A Fine Balance? Value-Relations, Post-capitalism and Forest Conservation – A Case from India', *Conservation and Society*, September (2023) 1–12.

41 Via Campesina, 'Small Scale Sustainable Farmers are Cooling Down the Earth', Online List 2007: via-info-en@googlegroups.com; Media Release: 'Call to Mobilise for a Cool Planet, Copenhagen', December 2009: http://www.viacampesina.org.

42 Erosion, Technology, Convergence (ETC), 'Civil Society Declaration on Technology and Precaution at COP15 in Copenhagen', 2009: https://etcgroup.org; 'What is Geoengineering?': http://www.geoengineering.monitor.org.

43 James Goodman, 'Direct Action on Climate Change', Left Renewal Conference, University of Technology, Sydney, May 2010: www.search.org.au/projects/from-global-crisis-to-green-future-australian-left-renewal-conference.

44 Rising Tide, 'The Climate Movement is Dead: Long Live the Climate Movement', 2010: www.risingtidenorthamerica.org.

45 ETC, 2009.

46 Ariel Salleh (ed.), *Eco-Sufficiency and Global Justice: Women Write Political Ecology.* London: Pluto Press, 2009.

47 People's Protocol on Climate Change, 2008; Patrick Bond, 'Climate Justice Action', *ZNet*, 24 October 2009: www.zmag.org/zspace/commentaries/4023.

48 Walden Bello, *Deglobalization*. London: Zed Books, 2004.

49 Third World Network (TWN), 'WIPO: Traditional Knowledge Committee Ends with Uncertainty Over Its Future', 2009: www.twnside.org.sg.

Chapter 8

Adapting parts of the 'Editorial: A Sociological Reflection on the Complexities of Climate Change Research', *International Journal of Water*, 5 (2010) 285–97.

1 'Fluorinated Refrigerants': https://www.washingtonpost.com/climate-environment/2021/02/15/these-gases-your-grocerys-freezer-are-fueling-climate-change-biden-wants-fix-that/.

2 Medea Benjamin and Nicolas Davies, 'US Militarism's Toxic Impact on Climate Policy', *Other News*, 22 September 2021.

3 Louise Sales, 'Marine Cloud Brightening: A Fossil Fuel Industry Smokescreen?', *Chain Reaction*, 140 (2021) 29–30.

4 World Rainforest Movement (WRM), 'What Do Forests Have To Do With Climate Change, Carbon Markets and REDD?', 10 May 2017: http:// wrm.org.uy/books-and-briefings/what-do-forests-have-to-do-with-climate-change-carbon- markets-and-redd/.

5 Friends of the Earth International (FOEI), 'Media Release', 26 May 2021: http://www.foei.org/features/historic-victory-judge-forces-shell-to-drastically-reduce-co$_2$-emissions.

6 Fiona Harvey and Terry Macalister, 'BP Study Predicts Greenhouse Emissions Will Rise By Almost A Third in 20 Years', *The Guardian*, 15 January 2014: http://www.theguardian.com/business/2014/jan/16/bp-predicts-greenhouse-emissions-rise-third.

7   Larry Lohmann, 'The Endless Algebra of Climate Markets', *Capitalism Nature Socialism*, 22 (2011) 93–116; Larry Lohmann, 'White Climate, White Energy: A Time for Movement Reflection?' *Social Anthropology*, 21 March 2021: https://doi.org/10.1111/1469-8676.12995.

8   Michal Kravcik, 'Learn More at the Protection Against Floods, Droughts and Climate Change Seminar', Kosice, Slovakia, 25 February 2010.

9   Clive McAlpine, J. Ryan, L. Seabrook, S. Thomas, P. Dargusch, J. Syktus, R. Pielke Sr., A. Etter, P. Fearnside, and W. Laurence, 'More than CO_2: A Broader Paradigm for Managing Climate Change and Variability to Avoid Ecosystem Collapse', *Current Opinion in Environmental Sustainability*, 2 (2010) 334–46.

10 Ilya Prigogine, *From Being to Becoming: Time and Complexity in the Physical Sciences*. New York: Freeman, 1981.

11 Beyond Zero Emissions, *Zero Carbon Australia: Stationary Energy Plan*. Melbourne: University of Melbourne, Energy Research Institute, 2010: www.beyondzeroemissions.org.

12 Global Environment Facility (GEF), 'Implementation of the Poznan Strategic Programme on Technology Transfer', 2010: www.thegef.com/gef/ccpublist.

13 Ivan Illich, *Energy and Equity*. New York: Boyars, 1977; Richard York and Eugene Rosa, 'Key Challenges to Ecological Modernization Theory', *Organization & Environment*, 16 (2003) 273–88.

14 Kurt Rommetveit, Silvio Funtowicz, and Roger Strand, 'Knowledge, Democracy and Action in Response to Climate Change' in R. Bhaskar, C. Frank, K. Hoyer, P. Naess, and J. Parker (eds.), *Interdisciplinarity and Climate Change*. London: Routledge, 2010, p. 56.

15 Thomas Kuhn, *The Structure of Scientific Revolutions*. London, Routledge, 1961.

16 IPCC, 'Climate Change: Synthesis Report', 2007: http://www.ipcc.ch/ cited in Ann Henderson-Sellers, 'The IPCC Report: What the Lead Authors Really Think', *Talking Point*, 2008: environmentalresearchweb.com.

17 Jan Pokorný, J. Brom, J. Čermák, P. Hesslerová, H. Huryna, N. Nadezhdina, and A. Rejšková. 'Solar Energy Dissipation and Temperature Control by Water and Plants', *International Journal of Water*, 5 (2010) 311–36.

18 Wilhelm Ripl, 'Losing Fertile Matter to the Sea: How Landscape Entropy Affects Climate', *International Journal of Water*, 5 (2010) 353–64.

19 Anastassia Makarieva and Victor Gorshkov, 'The Biotic Pump: Condensation, Atmospheric Dynamics and Climate', *International Journal of Water*, 5 (2010) 365–85.

20 Sheila Jasanoff, 'Testing Time for Climate Science', *Science*, 328 (2010) 695–96: www.sciencemag.org.

21 Stephen Toulmin, *Return to Reason*. Cambridge, MA: Harvard University Press, 2003, pp. 15–16.

22 Ana Isla, 'Who Pays for Kyoto Protocol?' in A. Salleh (ed.), *Eco-Sufficiency and Global Justice: Women Write Political Ecology*. London: Pluto Press, 2009; Ana Isla (ed.), *Climate Chaos: Ecofeminism and the Land Question*. Toronto: Innana, 2019.

23 Via Campesina, 'Thousands of Cancuns for Climate Justice!', 2010: viacampesina@viacampesina.org.

24 ETC Group, *Civil Society Declaration on Technology and Precaution at COP15 in Copenhagen*, 2009: https://etcgroup.org.

25 Vandana Shiva, *Staying Alive: Women, Ecology and Development*. London: Zed Books, 1989; Ariel Salleh, 'Is Our Sustainability Science Racist?' Ockham's Razor Program, Australian Broadcasting Corporation, Radio National, 4 October 2009: www.abc.net.au/RN/ockhamsrazor/stories/2009/2702106.htm.

26 Water for the People Network, *People's Water Resource Management Strategies*. Quezon City: IBON, 2009.

27 Juraj Kohutiar and Michal Kravcik, 'Water for an Integrative Climate Paradigm', *International Journal of Water*, 5 (2010) 298–310.

28 Duane Norris and Peter Andrews, 'Re-Coupling the Carbon and Water Cycles by Natural Sequence Farming', *International Journal of Water*, 5 (2010) 386–96.

29 Inter-Governmental Panel on Climate Change, *Sixth Synthesis Report*. Interlaken: United Nations, 2023; Brad Plumer, 'World Has Less Than a Decade to Stop Catastrophic Warming, UN Panel Says', *New York Times*, 20 March 2023: https://www.nytimes.com/2023/03/20/climate/global-warming-ipcc-earth.html.

30 Alex Bainbridge, 'Australia Shouldn't Wait to Phase Out Fossil Fuels', *Green Left*, 19 December 2023.

31 UNEP Foresight Brief, 'Working with Plants, Soils and Water to Cool the Climate and Rehydrate Earth's Landscapes', 2021: https://wedocs.unep.org/bitstream/handle/20.500.11822/36619/FB025.pdf.

32 Pablo Solon, 'Meeting on Climate Change, COP 27 and the Earth Assemblies Proposal', 24 August 2022: http://www.defenddemocracy.press/tens-of-thousands-take-to-the-streets-in-nationwide-protests-over-cost-of-living/.

Chapter 9

Adapting 'A Regenerative Ethic for a Gender Just Transition', *Institute for Global Development Magazine*, 28 January 2021: https://www.igd.unsw.edu.au/our-initiatives/equity-social-justice/gender-and-just-transitions.

1 Ariel Salleh, 'Ecofeminism' in Clive Spash (ed.), *Routledge Handbook of Ecological Economics*. Oxford & New York: Routledge, 2017, pp. 48–56.

2 Mary Mellor, 'Ecofeminist Political Economy' in Ariel Salleh (ed.), *Eco-Sufficiency and Global Justice: Women Write Political Ecology*. London: Pluto Press, 2009.

3 Jason Hickel, 'The Problem with Saving the World', *Jacobin Magazine*, 8 August 2015.

4 Patrick Murphy, 'The Ecofeminist Subsistence Perspective Revisited in an Age of Land Grabs', *Feminismo/s*, 22 (2013) 205–24; Ashish Kothari, Ariel Salleh, Arturo Escobar, Federico Demaria, and Alberto Acosta (eds.), *Pluriverse: A Post-Development Dictionary*. New Delhi: Tulika and AuthorsUpFront, 2019.

Chapter 10

1 Jeremiás Máté Balogh and Attila Jámbor, 'The Environmental Impacts of Agricultural Trade: A Systematic Literature Review', *Sustainability,* 12 (2020) 11–52; Bethany Reilly, 'Tens of Thousands Take to the Streets', *Defend Democracy Press*, 3 October 2022: http://www.defenddemocracy.press/tens-of-thousands-take-to-the-streets-in-nationwide-protests-over-cost-of-living/.

2 Xin Song, Guanqui Li, Ronnie Vernooy, and Yiching Song, 'Community Seed Banks in China: Achievements, Challenges and Prospects', *Agroecology and Ecosystem Services: Frontiers in Sustainable Food Systems,* 5 (2021): Article 630400. https://doi.org/10.3389/fsufs.2021.630400.

3 Little Donkey Farm; Green Ground Eco-Tech Centre: http://littledonkeyfarm.com/; Wen Tie Jun, Zhou Changyong, and Lau Kin Chi (eds.), *Sustainability and Rural Reconstruction*. Beijing: Agricultural University Press, 2015; Kin Chi Lau and Remy Herrera (eds.), *The Struggle for Food Sovereignty*. London: Pluto Press, 2015.

4 Sit Tsui, 'Rural Reconstruction' in A. Kothari, A. Salleh, A. Escobar, F. Demaria, and A. Acosta (eds.), *Pluriverse: A Post-Development Dictionary*. New Delhi: Tulika and AuthorsUpFront, 2019.

5 Song et al., 2021, p. 4.

6 Conference Program, 'Women, Biodiversity and Sustainable Food Systems in a Changing Climate', Academy of Science, Beijing, November 2019.

7 Song et al., 2021, p. 12.

8 Michel Pimbert, Nina Moeller, Brajesh Singh, and Colin Anderson, 'Agroecology' in *Oxford Research Encyclopedia of Anthropology*, p. 9: https://doi.org/10.1093/acrefore/9780190854584.013.298; Elna Tulus, Review: 'William I Robinson's Can Global

Capitalism Endure', 21 Match 2023: https://www.ppesydney.net/william-i-robinson-can-global-capitalism-endure/.

9 Indigenous Peoples' Global Summit on Climate Change, *The Anchorage Declaration*, 24 April 2009: envirosoc@listserve.brown.edu, 15 May 2009; Leonardo Figueroa-Helland, Cassidy Thomas, and Abigail Pérez Aguilera, 'Decolonizing Food Systems: Food Sovereignty, Indigenous Revitalization, and Agroecology as Counter-Hegemonic Movements', *Perspectives on Global Development and Technology*, 17 (2018) 173–201.

10 Belinda Lopez, 'When Rudd Sticks', *New Matilda*, 17 June 2008: www.newmatilda.com.

11 Mathias Wackernagel and William Rees, *Our Ecological Footprint*. Gabriola Island, BC: New Society, 1996: www.footprintnetwork.org.

12 Nicholas Georgescu-Roegen, *The Entropy Law and the Economic Process*. Cambridge, MA: Harvard University Press, 1971.

13 Ariel Salleh, *Ecofeminism as Politics: Nature, Marx, and the Postmodern*. London: Zed Books, 1997/2017.

14 John Bellamy Foster, *Marx's Ecology*. New York: Monthly Review, 2000; Peter Dickens, *Reconstructing Nature*. London: Routledge, 1995; Silvia Federici, *Caliban and the Witch*. New York: Autonomedia, 2004; Salleh, 1997/2017.

15 Wikipedia, 'Ecological Economics', *Wikipedia: The Free Encyclopedia*. www.wikipedia.

16 Herman Daly, Jon Erickson, and Joshua Farley, *Ecological Economics: A Workbook for Problem Based Learning*. Washington: Island Press, 2005.

17 FAO, 'International Conference on Organic Agriculture and Food Security', Rome, May, 2007; Catherine Badgley, J. Moghtader, E. Quintero, E. Zakem, M. Jahi Chappell, K. Aviles Vazquez, A, Samulon, and I. Perfecto, 'Organic Agriculture and the Global Food Supply', *Renewable Agriculture and Food System*s, 22 (2007) 86–108, p. 86.

18 Vandana Shiva, *Staying Alive: Women, Ecology, and Development*. London: Zed Books, 1989, p. 45; Veronika Bennhold-Thomsen and Maria Mies, *The Subsistence Perspective*. London: Zed Books, 1999.

19 Deborah Rose, *Nourishing Terrains*. Canberra: Australian Heritage Commission, 1996; John Gowdy (ed.), *Limited Wants, Unlimited Means*. Washington: Island Press, 1998.

20 Kirkpatrick Sale, *Dwellers in the Land: The Bioregional Vison*. San Francisco, CA: Sierra Club, 1985/2000, p. 42.

21 Barbara Adam, *Timescapes of Modernity,* London: Routledge, 1998.

22 Ariel Salleh (ed.), *Eco-Sufficiency and Global Justice: Women Write Political Ecology*. London: Pluto Press, 2009, pp. 302–3.

23 Ariel Salleh, 'Global Alternatives and the Meta-Industrial Class' in R. Albritton, J. Bell, S. Bell, and R. Westra (eds.), *New Socialisms: Futures Beyond Globalization*. London: Routledge, 2004.

24 Elizabeth Mpofu and Edgardo Garcia, 'Here is Why we are Boycotting the UN Food Systems Summit', *Al Jazeera*, 27 July 2021: https://www.other-news.info/here-is-why-we-are-boycotting-the-un-food-systems-summit/.

Chapter 11

Published online as 'Rio+20 and the Green Economy: Technocrats, Meta-industrials, WSF, and Occupy', *ZNet*, 31 March 2012: www.zcommunications.org/rio-20-and-the-green-economy-technocrats-meta-industrials-wsf-and-occupy; French Translation: 'Rio+20 et L'économie verte, les technocrates, les meta-industriels, le Forum Social Mondial et Occupy', *La Decouverte / Mouvements*, 70 (2012) 83–98.

1 UNCSD-Rio+20, *The Future We Want*, 30 January 2012: www.uncsd2012.org/rio20; Business Action for Sustainable Development, 'Media Release on the Zero Draft': 30 January 2012: www.UNEP.org.

2 Mujeres Manifesto, 'First Continental Summit of Indigenous Women', *Lucha Indigena*, 34 (2009); Isagani Serrano, *What Sustainable Development?* Manila: Philippine Rural Reconstruction Movement, 2011.

3 Ariel Salleh, 'Global Alternatives and the Meta-Industrial Class' in R. Albritton, J. Bell, S. Bell, and R. Westra (eds.), *New Socialisms: Futures Beyond Globalization*. London: Routledge, 2004; Ariel Salleh, 'From Metabolic Rift to Metabolic Value: Reflections on Environmental Sociology and the Alternative Globalization Movement', *Organization & Environment*, 23 (2010) 205–19.

4 International Institute for Sustainable Development (IISD), *Earth Negotiations Bulletin*, 22 February 2012: www.iisd.ca/unepgc/unepss12/#3.

5 La Via Campesina, 'Call to action: Reclaiming our Future: Rio+20 and Beyond', 16 February 2012: via-info-en@googlegroups.com. Tom Mertes (ed.), *The Movement of Movements*. London: Verso, 2004; Jackie Smith, Marina Karides, Marc Becker, Dorval Brunelle, Christopher Chase-Dunn, and Donatella Della Porta (eds.), *Global Democracy and the World Social Forums*. London: Routledge, 2014.

6 Dialogue Platform of the Thematic Social Forum, *Another Future is Possible: Come to Re-Invent the World at Rio+20*, Porto Alegre, 24 January 2012. Christophe Aguiton and Nicolas Haeringer, 'Occupy the Left: A Few Thoughts on Current Movements and the Left', 6 March 2012: http://openfsm.net/projects/occupy-and-wsf/occupy-wsfindex; Marina Sitrin (ed.), *Horizontalism: Voices of Popular Power in Argentina*. Oakland, CA: AK Press, 2006.

7 ETC, 'Who Will Control the Green Economy? Building the Peoples Summit Rio+20', *Rio+20 Portal,* 17 December 2011: info@forums.rio20.net, italics added.

8 Worldwatch Institute, *Toward a Transatlantic Green New Deal: Tackling the Climate and Economic Crises.* Brussels: Heinrich-Boell-Stiftung, 2009.

9 Ariel Salleh (ed.), *Eco Sufficiency and Global Justice: Women Write Political Ecology.* London: Pluto Press, 2009.

10 IISD Reporting Service, 'Commission on the Status of Women Focuses on Empowerment of Rural Women and their Role in Sustainable Development', 27 February 2012: www.iisd.ca/.

11 Gigi Francisco and Peggy Antrobus, 'Mainstreaming Trade and Millennium Development Goals?' in Salleh (ed.), 2009.

12 New Economics Foundation, *A Green New Deal: Joined Up Policies.* London: NEF, 2008.

13 IISD Reporting Service, 'Briefing Note on the Thirteenth Major Groups and Stakeholders Forum', 20 February 2012: www.iisd.ca/unepgc/unepss12/gmgsf13.

14 Felipe Calderon, 'G20 Finance Ministers and Chancellors Discuss Green Growth', 26 February 2012: www.g20.org/newsroom/.

15 Environmental Justice Organizations, Liabilities and Trade: www.ejolt.org. Chris Lang, 'Finance for Biodiversity is a New Face for Capitalism: Sign on Letter to CBD from Accion Ecologica', 27 January 2012: www.wp.me/pll98-2Ss.

16 George Russell, 'Obama administration sees Rio+20 Summit in June as Festival of Global Greenness', *Fox News,* 4 February 2012: www.FoxNews.com.

17 Business Action for Sustainable Development (BASD), 'Media Release on the Zero Draft': 30 January 2012: www.UNEP.org, italics added.

18 Dialogue Platform of the Thematic Social Forum, 2012.

19 Maureen Taylor and George Friday, 'Occupy Wall Street and the US Social Forum Movement: Local and National Perspectives', 17 February 2012: http://www.ussf2010.org/node/372.

20 M. Steisslinger, 'Occupy, the World Social Forum and the Commons', 13 March 2012: http://thefutureofoccupy.org/2012/03/13/occupy-the-world-social-forum-and-the-commons-social-movements-learning-from-each-other/; Homad Hosseini, 'Transversality in Diversity: Experiencing Networks of Confusion and Convergence in the World Social Forums', *International and Multidisciplinary Journal of Social Sciences,* 4 (2015), 54–87.

21 Stephen Lerner, 'Horizontal Meets Vertical', *The Nation,* 2 April 2012, p. 20.

22 Chico Whitaker, 'New Perspectives in the WSF Process', *Alternatives International,* 30 January 2012: www.alterinter.org/article3745.html?lang=fr.

23 Dialogue Platform of the Thematic Social Forum, 2012.

24 Caroline Lucas, 'A Political View' in C. Farrell, A. Green, S Knights, and W. Skeaping (eds.), *This Is Not a Drill: An Extinction Rebellion Handbook*. London: Penguin, 2019.

Chapter 12

Published online by the Institute for Social Analysis, Rosa Luxemburg Stiftung, Berlin as 'A Critical Feminist Reading of the Green New Deal': ifg.rosalux.de/files/2010/04/Salleh-RLS-GND.pdf; Adapted as 'Green New Deal or Globalisation Lite?', *Arena Magazine*, 105 (2010) 15–19.

1 Rosa Luxemburg, *The Accumulation of Capital*. New York: Monthly Review Press, 1913/1968.

2 Ariel Salleh, 'Ecofeminism' in C. Spash (ed.), *Routledge Handbook of Ecological Economics*. London: Routledge, 2017.

3 Vijay Kolinjivadi and Ashish Kothari, 'No Harm Here is Still Harm There: The Green New Deal and the Global South', Parts I, 2021: https://beyonddevelopment.net/category/transformations/?print=print-search; Vijay Kolinjivadi and Ashish Kothari, Part II 2021: https://www.jamhoor.org/read/2020/5/20/no-harm-here-is-still-harm-there-looking-at-the-green-new-deal-from-the-global-south.

4 New Economics Foundation, *A Green New Deal: Joined Up Policies*. London: NEF, 2008; Tim Jackson, *Prosperity Without Growth? The Transition to A Sustainable Economy*. London: Sustainable Development Commission, 2009.

5 Tim Jackson, *The Politics of Happiness: A NEF Discussion Paper*. London: New Economics Foundation, 2003 cited in Louise Crossley, 'Discussion Paper: Green New Deal: Ecology, Economy, Democracy', Green Institute, Melbourne, May 2009, p. 6.

6 United Nations Environment Programme (UNEP), *Global Green New Deal*, London/Nairobi, UNEP, 22 October 2008.

7 International Conference on Organic Agriculture and Food Security, Rome, May 2007; Bill Mollison, *Permaculture: A Designer's Manual*. Tyalgum, NSW: Tagari, 1988.

8 New South Wales Department of Climate and Conservation: http://www.environment.nsw.gov.au/threatspec/publicconsult.htm.

9 Annie James and Neena Pathak Broome, 'A Fine Balance? Value-Relations, Post-Capitalism and Forest Conservation – A Case from India', *Conservation and Society*, September (2023) 1–12.

10 UNEP, 2008.

11 Worldwatch Institute, *Toward a Transatlantic Green New Deal: Tackling the Climate and Economic Crisis*. Brussels: Heinrich-Boell-Stiftung, 2009.

12 Worldwatch Institute, p. 10.

13 Worldwatch Institute, p. 21; italics added.

14 Worldwatch Institute, p. 5.

15 Worldwatch Institute, p. 11.

16 South Peoples Historical, 'Social Ecological Debt Creditors Alliance', *Quito Statement*, 22 August 2007: www.accionecologica.org.

17 Worldwatch Institute, p. 14.

18 Worldwatch Institute, pp. 13–14; italics added.

19 Worldwatch Institute, p. 19.

20 Worldwatch Institute, p. 22.

21 *Joint Statement: Towards a Green New Deal: Economic Stimulus and Policy Action for the Double Crunch*: www.acfonline.au.org.

22 AGIC: www.agic.net.au; National Infrastructure Conference: Building and Investing in the Future, Sydney, 1–2 April 2009.

23 *Joint Statement*, p. 3.

24 Australian Council of Trade Unions, *Green Gold Rush: The Future of Australia's Green Collar Economy*. Melbourne: ACTU, 2008.

25 Christine Milne, *Re-Energising Australia*. Canberra: Australian Green Party, 2007; Crossley, 2009; Frances Flanagan, 'Horizons of National Responsibility: Law and the Protection of Life on a Fossil Rich Continent', Hancock Lecture, Australian Academy of the Humanities, Sydney, 31 May 2022.

26 Marilyn Waring, 'Policy and the Measure of Woman: UNSNA, ISEW, HDI, and GPI' in Ariel Salleh (ed.), *Eco-Sufficiency and Global Justice: Women Write Political Ecology*. London: Pluto Press, 2009.

27 Alexandria Ocasio-Cortez, 'Recognizing the duty of the Federal Government to Create a Green New Deal', House of Representatives: http://ocasio-cortez.house.gov/gnd; Alexandria Ocasio-Cortez, Green New Deal US: http://assets.documentcloud.org/documents/5729033/Greem-New-Deal-FINAL.pdf

28 John Bellamy Foster, *Capitalism in the Anthropocene: Ecological Ruin or Ecological Revolution*. New York: Monthly Review Press, 2022.

29 Anna Sturman and Natasha Heenan, 'Configuring the Green New Deal', *The Economic and Labour Relations Review*, 32 (2021) 149–54, p. 152.

30 Sustainable Europe Research Institute (SERI), UN University, and Finland Futures Research Centre, *Environment and Innovation*, Vienna: SERI, 2006.

31 Sofia Scasserra and Carolina Martinez Elebi, Transnational Institute, 6 October 2021: http://www.tni.org/en/publication/digital-colonialism.

32 David Adler and Pawel Wargan, 'The EU's Green Deal the Betrayal of a Generation: A Strategy to Fight Back'; and Birgit Mahnkopf, 'On the Political Economy of the Ecological Crisis' in W. Baier, E. Canepa, and H. Golemis, *Transform! Yearbook 2021 Capitalism's Deadly Threat*. London: Merlin Press, 2021.

33 Ted Trainer, *Abandon Affluence!* London: Zed Books, 1985; Ted Trainer, 'A (Friendly) Critique of the Degrowth Movement', *The Simpler Way*, January 2024: https://thesimplerway.info/DEGROWTHCRIT.pdf.

34 Stefania Barca, 'The Labor(s) of Degrowth', *Capitalism Nature Socialism*, 30 (2019) 207–16; Joan Martinez-Alier, *The Environmentalism of the Poor: A Study of Ecological Conflicts and Valuation*. Cheltenham, UK: Edward Elgar, 2002.

35 Damian White, Alan Rudy, and Brian Gareau, *Environments, Natures and Social Theory: Towards a Critical Hybridity*. London: Palgrave, 2016.

36 Ariel Salleh, 'From Metabolic Rift to Metabolic Value: Reflections on Environmental Sociology and the Alternative Globalization Movement', *Organization & Environment*, 23 (2010) 205–19; John Bellamy Foster, *Marx's Ecology: Materialism and Nature*. New York: Monthly Review Press, 2000.

37 Stefania Barca, 'Labour and the Ecological Crisis: The Eco-Modernist Dilemma in Western Marxism(s) 1970s–2000s', *Geoforum*, 98 (2019) 226–35, p. 4.

38 DiEM25, The Green New Deal for Europe: http://report.gndforeurope.com.cms/wp-content/uploads/2020/01/Blueprint-for-Europes-Just-Transition-2nd-Ed.pdf.

39 Democratic Socialists of America, 'DSA's Green New Deal Principles': http://ecosocialists.dsusa.org/2019/02/28/gnd-principles/.

40 DSA: http://ecosocialists.dsusa.org/2019/02/28/gnd-principles/.

41 DSA: http://ecosocialists.dsusa.org/2019/02/28/gnd-principles/.

42 The Red Nation, *The Red Deal: Indigenous Action to Save Our Earth*. New York: Common Notions, 2021, cited in Simon Butler and Ian Angus, 'The Red Deal to Save Earth', *Green Left*, 7 September 2021: www.greenleft.org.au.

43 Leonardo Figueroa-Helland, notes for the book *Indigenous Resurgence beyond Anthropocene Collapse*, forthcoming.

44 Julia Marti Comas, 'Ecofeminist Review of the Proposals for a Green New Deal', *Transform!*, 2020, p. 10.

45 Bhumika Muchhala, 'Feminist Economic Justice for People and Planet', *WEDO Issue Brief*, May 2021.

46 Bhumika Muchhala, 'Towards a Decolonial and Feminist Global Green New Deal', Rosa-Luxemburg Stiftung: https://www.rosalux.de/en/news/id/43146/towards-a-decolonial-and-feminist-global-green-new-deal.

Chapter 13

Lecture delivered at Roskilde University as 'A Post-Development Reading of Climate and the SDGs'; published in Danish as 'En "post-development"- forstaelse af klima, vand og baeredygtigheds malene', *Clarte*, 30 (2016) 10–18.

1. UN Sustainable Development Knowledge Platform, *Transforming Our World: The 2030 Agenda for Sustainable Development*, 2015: www.sustainabledevelopment.un.otg/post2015/transformingourworld/publications.

2. Peggy Antrobus, 'Mainstreaming Trade and Millennium Development Goals?' in Ariel Salleh (ed.), *Eco-Sufficiency and Global Justice: Women Write Political Ecology*. London: Pluto Press, 2009.

3. Jason Hickel, 'The Problem with Saving the World', *Jacobin Magazine*, 8 August 2015, p. 3.

4. Ariel Salleh, 'Green Economy or Green Utopia? Rio+20 and the Reproductive Labor Class', *Journal of World-Systems Research*, 18 (2012) 141–45.

5. Global Redesign Initiative, 2012: www.qatarconferences.org/economic/world/GRI_Executive_Summary.pdf.

6. Ariel Salleh, 'Neoliberalism, Scientism, and Earth System Governance' in R. Bryant (ed.), *International Handbook of Political Ecology*. Cheltenham: Elgar, 2015.

7. UN Sustainable Development Knowledge Platform, 2015, para 48.

8. James O'Connor, *Natural Causes*. New York: Guilford, 1998.

9. Hickel, 2015, p. 3.

10. David Harvey, *Spaces of Global Capitalism*. New York: Routledge, 2006, p. 96.

11. William Easterly, 'SDGs: Senseless, Dreamy, Garbled', *Foreign Policy*, 2015: http://foreignpolicy.com/2015/09/28/the-sdgs-are-utopian-and-worthless-mdgs-development-rise-of-the-rest/.

12. Danny Chivers and Jess Worth, 'Paris Deal: Epic Fail on a Planetary Scale', *New Internationalist*, 12 December 2015. People's Climate Test: http://peoplestestonclimate.org.

13. Patrick Bond, 'Can Climate Activists' "Movement Below" Transcend Negotiators' "Paralysis Above"?', *Journal of World-Systems Research*, 21 (2015) 250–70; Patrick Bond, 'The Case for Ecosocialism in the Face of the Worsening Climate Crisis', *Science & Society*, 86 (2022) 485–515.

14. Ashish Kothari, Ariel Salleh, Arturo Escobar, Federico Demaria, and Alberto Acosta (eds.), 'Introduction' in *Pluriverse: A Post-Development Dictionary*. New Delhi: Tulika & AuthorsUpFront, 2019, pp. xxvi–xxvii.

15 John Clark, *The Tragedy of Common Sense*. Regina, SK: Changing Suns Press, 2016.

16 Martin Winiecki and Leila Dregger, 'Water: The Missing Link for Solving Climate Change', *Terra Nova Voice*, 28 November 2015, p. 2.

17 Ariel Salleh, 'Editorial: A Sociological Reflection on the Complexities of Climate Change Research', *International Journal of Water*, 5 (2010) 285–97.

18 Via Campesina, 'Small Scale Sustainable Farmers are Cooling Down the Earth', Jakarta, Via Campesina Views, 2009: https://viacampesina.org/en/wp-content/uploads/sites/2/2010/03/Small-Scale-Farmers-Cool-Down-EarthEN.pdf.

19 UN Sustainable Development Knowledge Platform, 2015.

20 Maude Barlow and Tony Clarke, 'Water Privatization', Global Policy Forum, 2004: www.globalpolicy.org.

21 Conference on Post-2015 Sustainable Development Goals: Towards a New Social Contract, University of Rheims, June 2013.

22 Aurora Portal, Lourdes Contreras, and Rosa Rivero, 'Strike Against the AntaKori Mining Project: Peruvian Women and Communities Resist', *Capire: Newsletter of the World March of Women*, 26 March 2023: https://capiremov.org/en/experience/strike-against-the-antakori-mining-project/.

Chapter 14

1 Miriam Lang, 'Peculiarities of the State in the Periphery of the Modern Colonial World System', paper presented at the Degrowth Conference, Vienna, 29 May 2020.

2 Rob Wallace, *Big Farms Make Big Flu*. New York: Monthly Review Press, 2016; Yinon Bar-On, Rob Phillips and Ron Milo, 'The Bio-Mass Distribution on Earth', *PNAS*, 19 June 2018.

3 John Hinkson, 'A High Tech Pandemic?', *Arena*, 2 (2020) 62–71.

4 Countercurrents Collective, 15 April 2020: https://countercurrents.org/2020/04/governments-are-using-coronavirus-to-build-the-architecture-of-oppression-warns-edward-snowden/.

5 Vandana Shiva, 'Gates to the Global Empire', *Navdanya International*, 14 October 2020: https://navdanyainternational.org/bill-gates-philanthro-capitalist-empire-puts-the-future-of-our-planet-at-stake.

6 Anon, 'Ed-Tech Companies Exploit Crisis', *Education: Journal of the New South Wales Teachers Federation*, 101 (2020) 19–20; Ben Williamson and Anna Hogan, 'Commercialisation and privatisation in/of education in the context of Covid-19',

Education International: ei-ie.org/en/detail/16858/wdtech-pandemic-shock-new-ei-research-launched-on-covid-19-education-commercialisation.

7   Deane Neubauer, 'Public Health 4.0 in the Emergent Climate of Global Transformation' in H. Hosseini, J. Goodman, S. Motta and B. Gills (eds.), *The Routledge Handbook of Transformative Global Studies*. London: Routledge, 2021.

8   Klaus Schwab, *The Fourth Industrial Revolution*. New York: Currency, 2016, p. 1.

9   Marc Dion, Philip Abdelmalik and Abla Mawudeku, 'Big data and the global public health intelligennce network', *Big Data and GPHIN CCRR*, (2015) 41–9.

10 Yanis Varoufakis, 'Techno-Feudalism Is Taking Over', *The Other News*, 7 July 2021.

11 Shoshana Zuboff, *The Age of Surveillance Capitalism*. New York: Hachette, 2019, p. 83.

12 Vandana Shiva, *Oneness vs the 1%: Shattering Illusions, Seeding Freedom*. North Geelong, VIC: Spinifex, 2018.

13 Klaus Schwab, 'The Fourth Industrial Revolution and Why You Should Care', *Salesforce Blog*, 2 July 2020.

14 Naomi Klein, 'The Screen New Deal', *The Intercept*, 8 May 2020: https://www.democracynow.org/2020/5/13/naomi_klein_coronavirus_tech_privacy_surveillance.

15 Erosion, Technology, Convergence (ETC): https://etcgroup.org/mission; Filippa Lentzos and Guy Reeves, 'Scientists are working on vaccines that spread like a disease', *Bulletin of the Atomic Scientists*, 8 September 2020: https://thebulletin.org/2020/09/scientists-are-working-on-vaccines-that-spread-like-a-disease/.

16 Renate Klein, 'Reproductive Engineering' in A. Kothari, A. Salleh, A. Escobar, F. Demaria, A. Acosta (eds.), *Pluriverse: A Post-Development Dictionary*. New Delhi: Tulika/AuthorsUpFront, 2019.

17 Vishwas Satgar and Ruth Ntloketse (eds.), *Emancipatory Feminism in the Time of Covid-19: Transformative Resistance and Social Reproduction*. Johannesburg: Wits University Press, 2023.

18 Tim Shorrock, 'How Private Contractors Have Created a Shadow NSA', *The Nation*, 27, 2015: https://www.thenation.com/article/archive/how-private-contractors-have-created-shadow-nsa/.

19 Australian Cyber Security Centre, 'The ACSC Annual Cyber Threat Report July 2019 to June 2020': htto://sites/default/files/2020-09/ACSC-Annual-Cyber-Threat-Report-2019-20.pdf.

20 Shoshana Zuboff, Norma Moellers, David Murakami Wood and David Lyon, 'Surveillance Capitalism: An Interview with Shoshana Zuboff', *Surveillance & Society*, 17 (2019) 257–66, p. 261.

21 Just Net Coalition, 'More than 170 Civil Society Groups Worldwide Oppose Plans for a BigTech Dominated Body for Global Digital Governance': wwwjustnetcoalition.org;

Timothy Strom, 'Cybernetic Capitalism with Chinese Characteristics', *Arena*, 6 (2021) 24–31.

22 Oceania Radiofrequency Scientific Advisory Association (ORSAA), 2020: https://www.orsaa.org; Pri Bandara and David Carpenter, 'Planetary electromagnetic pollution: it is time to assess its impact' *The Lancet,* 514 (2018): www.thelancet.com/planetary-health.

23 PhoneGate Alert Archives, Press Releases: https://phonegatealert.org/cat/communiques/.

24 Mark Hertsgaard and Mark Dowie, 'How Big Wireless Made Us Think That Cell Phones Are Safe: A Special Investigation', *The Nation*, 29 March 2018: https://www.thenation.com/article/archive/how-big-wireless-made-us-think-that-cell-phones-are-safe-a-special-investigation.

25 United States Government, 'Secure 5G and Beyond Act of 2020': https://congress.gov/bill/116-congress/senate-bill/893.

26 European Council, 'Digitalisation for the Benefit of the Environment: Council Approves Conclusions': http://www.consilium.europa.eu/en/press/press-releases/2020/12/17...n-for-the-benefit-of-the-environment-council-approves-conclusions/.

27 Tom Butler, *Wireless Technologies and the Risk of Adverse Health Effects: A Retrospective Ethical Risk Analysis of Health and Safety Guidelines.* Working Paper, University College Cork, 2021; Barbara Koeppel, 'Wireless Hazards', *The Washington Spectator*, 28 December 2020: https://washingtonspectator.org/wireless-hazards/.

28 The Appeal to UN, WHO, EU, and Nation States, 2020: https://www.5gspaceappeal.org/the-appeal; Letter to Dr Tedros Adhanom Ghebreyesus, Director-General of the World Health Organization (WHO), *International Public Call for Protection from Non-ionizing Electromagnetic Field (EMF) Exposure,* signed by international experts in the field of bioelectromagnetics, January 2021.

29 Mark Steele, 'Fifth Generation (5G) Directed Energy Radiation Emissions in the Context of Contaminated Nanometal Covid-19 Vaccines with Graphite Ferrous Oxide Antennas', *Global Research*, 20 July 2022.

30 Yuval Harari, *Homo Deus: A Short History of Tomorrow.* London: Vintage, 2017.

31 Nancy Owano, 'CEET Report Nails Wireless as Energy Monster', 13 April 2013: https://phys.org/news/2013-04-CEET-wireless-energy-monster-html.

32 Environmental Health Trust, 2022: 'Climate Change and 5G': https://ehtrust.org/climate-change-and-5g/.

33 Jean-Marc Jancovici cited by environmental health trust: https://ehtrust.org/climate-change-and-5g/.

34 Mathew Barton, 'Smart tech's carbon footprint', *The Ecologist*, 30 April 2020.

35 Masha Borak, '5G Base Stations Use up to Three-and-a-Half Times More Energy than 4G Infrastructure', 27 August 2020: https://www.scmp.com/abacus/tech/article/3098964/5g-towers-are-consuming-lot-energy-so- china-unicom-putting-some-them; EMFacts: https://www.emfacts.com/2020/09/.

36 Charlotte Trueman, 'Why Data Centers Are the New Frontier in the Fight Against Climate Change', *Computerworld*, 10 August 2019: https://www.computerworld.com/article/3431148/why-data-centres-are-the-new-frontier-in-the-fight-against-climate-change.html.

37 Oliver Belcher, Patrick Bigger, Ben Neimark and Cara Kennelly, 'Hidden Carbon Costs of the "Everywhere War": Logistics, Geopolitical Ecology, and the Carbon Bootprint of the US Military', *Transactions of the Institute of British Geographers*, 45 (2020) 65–80.

38 Erosion, Technology, Convergence (ETC), 'Geopiracy: The Case Against Geoengineering, 2010: http://www.etcgroup.org/en/node/5217; Louise Sales, 'Geoengineers Test Risky Planetary Engineering Scheme in Australia: Experiment Defies 193 Country UN Moratorium', 11 May 2020.

39 Margarida Mendes, 'Oceano Livre: Coalition Against Deep Sea Mining', 2018: http://oceanolivre.org/en; Amalia Hart, 'Mining the Moon', *Cosmos Weekly*, 3 February 2023.

40 Hiroko Tabuchi and Brad Plumer, 'How Green Are Electric Vehicles?', *New York Times*, 28 October 2021: https://www.nytimes.com/2021/03/02/climate/electric-vehicles-environment.html.

41 Jonathan Marshall, James Goodman, Didar Zowghi and Francesca da Rimini, *Disorder and the Disinformation Society: The Social Dynamics of Information, Networks and Software*. London: Routledge, 2015, p. 14.

42 Ariel Salleh, 'Neoliberalism, Scientism and Earth System Governance' in R. Bryant (ed.), *The International Handbook of Political Ecology*. London: Elgar, 2015.

43 John Bellamy Foster, 'Marx's Theory of Metabolic Rift: Classical Foundations for Environmental Sociology', *American Journal of Sociology*, 105 (1999) 366–405.

44 Trueman, 2019, p. 4.

45 Shigeaki (Shey) Hakusui, 'Fixed Wireless Communications at 60GHz Unique Oxygen Absorption Properties', *RF Globalnet News*, 10 April 2001: https://www.rfglobalnet.com/doc/fixed-wireless-communications-at-60ghz-unique-0001.

46 Zebedee Nicholls and Tim Baxter, 'Climate Explained: Methane Is Short-Lived in the Atmosphere but Leaves Long-Term Damage', *The Conversation*, 9 September 2020: https://theconversation.com/climate-explained-methane-is-short-lived-in-the-atmosphere-but-leaves-long-term-damage-145040.

47 'Space Pollution': https://www.theverge.com/2019/6/28/19154142/spacex-starlink-60-satellites-communication-internet-constellation.

48 Rosalie Bertell, *Planet Earth the Latest Weapon of War*. London: Womens Press, 2000; Ariel Salleh, 'Ecosocialism, Gendered Imaginaries and the Informatic-Securitization Complex', *Capitalism Nature Socialism*, 25 (2014) 24–39.

49 Arthur Firstenberg, *The Invisible Rainbow: A History of Electricity and Life*. London: Chelsea Green, 2020.

50 Jan Zalasiewicz, M. Williams, C. Waters, A. Barnosky, J. Palmesino, A. Rönnskog, M. Edgeworth, C. Neal, A. Cearreta, E. Ellis, J. Grinevald, P. Haff, J. Ivar do Sul, C. Jeandel, R. Leinfelder, J. McNeill, E. Odada, N. Oreskes, S. Price, A. Revkin, W. Steffen, C. Summerhayes, D. Vidas, S. Wing and A. Wolfe. 'Scale and Diversity of the Physical Technosphere: A Geological Perspective', *The Anthropocene Review*, 2016, 1–14, p.3; italics added: DOI:10.1177/2053019616677743.

51 'A Nuclear Accident Made Three Mile Island Infamous: AI's needs may revive it', 10 July 2024: http://www.washingtonpost.com/business/2024/07/10/three-mile-island-nuclear-artificial-intelligence.

52 Peter Bloom and Alberto Acosta, 'Mining Infinity: Extractivism's Final Frontier', 2021: https://ecor.network/news/mining-infinity-extractivism-s-final-frontier/.

53 Elizabeth Steyn, 'Space Mining Is Not Science Fiction, and Canada Could Figure Prominently', *The Conversation*, 4 April 2021: https://theconversation.com/space-mining-is-not-science-fiction-and-canada-could-figure-prominently-155855.

54 Antonio Busalacchi, 'The Growing Risk of Space Weather', *Corporate Risk and Insurance*, 24 August 2018: https://www.insurancebusinessmag.com/us/risk-management/operations/the-growing-risk-of-space-weather-109672.aspx.

55 Sandra Burmeier, Reto Schneider, Philippe Brahin, *Swiss Re SONAR, Emerging Risk Insights*. Zurich: Corporate Real Estate & Logistics/Media Production, 2013.

56 American Physics Association (APA), 'Phone Radiation May Be Killing Insects: German study', 17 September 2020: https://phys.org/news/2020-09-mobile-insects-german.html.

57 Ariel Salleh, 'A Materialist Ecofeminist Reading of the Green Economy' in H. Hosseini, J. Goodman, S. Motta and B. Gills (eds.), *The Routledge Handbook of Transformative Global Studies*. London: Routledge, 2020.

58 Patrick Bond, 'Who Really "State-Captured" South Africa?' in E. Durojaye and G. Mirugi-Mukundi (eds.), *Exploring the Link between Poverty and Human Rights in Africa*. Pretoria: Pretoria University Law Press, 2020; Patrick Bond, 'The 4IR, from Sociological Critique to Social Resistance', 1 March 2021: wsmdiscuss@lists.opespaceforum.net.

Chapter 15

1 Secretary-General's High Level Panel on Digital Cooperation: https://www.un.org/en/sg-digital-cooperation-panel#.

2 People's Letter to Sima Sami Bahous, UN Women's Executive Director, 'UN Women's MOU with Blackrock', Circulated Online, 24 July 2022: www.focusweb.com.

3 Shoshana Zuboff, *The Age of Surveillance Capitalism*. New York: Hachette, 2019.

4 Caroline Criado Perez, *Invisible Women: Exposing Data Bias in a World Designed for Men*. London: Chatto & Windus, 2019.

5 Britt Baatjes, *Beware of the Bot: A Critical Perspective on the 4IR*. Port Elizabeth: Nelson Mandela University, CIPSET, 2020.

6 Jason Lewis (ed.), *Indigenous Protocol and Artificial Intelligence Position Paper*. Honolulu: The Initiative for Indigenous Futures and the Canadian Institute for Advanced Research (CIFAR), 2020: https://doi.org/10.11573/spectrum.library.concordia.ca.00986506.

7 Lewis, 2020, p. 42.

8 Universite de Montreal, *The Montreal Declaration for Responsible AI Development*, 2018: montrealdeclaration-responsibleai.com/the-declaration.

9 Lewis, 2020, p. 156; Pinar Tuzcu, 'Decoding the Cybaltern: Cybercolonialism and Postcolonial Intellectuals in the Digital Age', *Postcolonial Studies*, 24 (2021) 514–27.

10 Bruno Latour, *Reassembling the Social: An Introduction to Actor Network Theory*. Oxford: Oxford University Press, 2005; Donna Haraway, *Staying With the Trouble: Making Kin in the Chthulucene*. Durham, NC: Duke University Press, 2016.

11 Jason Lewis, Noelani Arista, Archer Pechawis, and Suzanne Kite, 'Making Kin with the Machines', *Journal of Design and Science,* 13 (2018) quoted in Lewis, 2020, p. 4.

12 Lewis, 2020, p. 153.

13 Lewis, 2020, p. 28; italics added.

14 Lewis, 2020, p. 66.

15 Karen Barad, *Meeting the Universe Halfway: Quantum Physics and the Entanglement of Matter and Meaning*. Durham, NC: Duke University Press, 2007; Jane Bennett, *Vibrant Matter: A Political Ecology of Things*. Durham, NC: Duke University Press, 2010.

16 Lewis, 2020, p. 72.

17 Kate Crawford and Vladan Joler, 'Anatomy of an AI System: The Amazon Echo As an Anatomical Map of Human Labor, Data and Planetary Resources', *AI Now Institute and Share Lab*, 2 September 2018.

18 Lewis, 2020, p. 76.

19 Lewis, 2020, p. 120.

20 Lewis, 2020, p. 151.

21 Alf Hornborg, *Global Ecology and Unequal Exchange: Fetishism in a Zero-Sum World*. London: Routledge, 2011.

22 Graham Brewer, 'Is Copyright law a 'Colonization of Knowledge?', *High Country News*, March, 2019.

23 Lewis, 2020, p. 21; J. Crabtree, 'Cambridge Analytica is an "Example of What Modern Day Colonialism Looks Like", Whistle Blower Says', *CNBC*, 27 March 2018.

24 Lewis, 2020, p. 104.

25 United Nations Year of Indigenous Languages: www.en.iyil2019.org.

26 Klaus Schwab, 'The Fourth Industrial Revolution: What it Means; How to Respond', *World Economic Forum*, 2016: www.weforum.org/agenda/2016/01/the-fourth-industrial-revolution-what-it-means-and-how-to-respond.

27 Ray Kurzweil, *The Singularity 2045*: http://www.singularity.com/.

28 Lewis, 2020, p. 200.

29 Toby Walsh et al., 'Report for the Australian Council of Learned Academies', 2019: acola.org/wp-content/uploads/2019/097/hs4_artificial-intelligence-report.pdf.

30 Whaanga in Lewis, 2020, p. 34.

31 Boaventura de Sousa Santos, *Epistemologies of the South: Justice Against Epistemicide*. Boulder, CO: Paradigm, 2014; Frantz Fanon, *Black Skin, White Masks*. New York: Glove Press, 1967; Linda Tuhiwai Smith, *Decolonizing Methodologies*. London: Zed Books, 1999.

32 Geek Girl Academy: 'What Would the Internet Look Like if More Women were Designing It?', 2020: https://girlgeekacademy.com/about/.

33 Guillaume Pitron, 'Toxic Secrets Behind Your Mobile Phone', *Daily Mail Online*, 7 February 2021: http://www.dailymail.co.uk/news/article-9179751/Toxic-secrets-mob.

34 Sabelo Ndlovu-Gatsheni, 'Global Economy of Knowledge in Transformative Global Studies' in H. Hosseini, J. Goodman, S. Motta, and B. Gills (eds.), *The Routledge Handbook of Transformative Global Studies*. London: Routledge, 2020.

35 Arturo Escobar, *Pluriversal Politics: The Real and the Possible*. Durham, NC: Duke University Press, 2017, pp. xvi–xvii; Marisol dela Cadena and Mario Blaser (eds.), *A World of Many Worlds*. Durham, NC: Duke University Press, 2018.

36 H. Lai, 'Research Summary: ELF-EMF/Static Field Neurological Effects Abstracts' in C. Sage and D. Carpenter (eds.), *BioInitiative Report: A Rationale for a Biologically-based Public Exposure Standard for Electromagnetic Fields (ELF and RF)*, 2020: https://bioinitiative.org.

37 UNEP-WCMS, *State of the World's Migratory Species*. Cambridge, UK: United Nations, 2024.

38 Ashish Kothari, Ariel Salleh, Arturo Escobar, Federico Demaria, and Alberto Acosta (eds.), *Pluriverse: A Post-Development Dictionary*. New Delhi: Tulika and AuthorsUpFront, 2019.

39 Michael Shellenburger and Ted Nordhaus, Breakthrough Institute: http//the breakthrough.org.

40 Jason Hickel and Giorgo Kallis, 'Is Green Growth Possible?', *New Political Economy*, 25 (2020) 469–86.

41 Vijay Kolinjivadi and Ashish Kothari, 'No Harm Here is Still Harm There: The Green New Deal and the Global South', Part I, 2021: https://beyonddevelopment.net/category/transformations/?print=print-search; Vijay Kolinjivadi and Ashish Kothari, Part II 2021: https://www.jamhoor.org/read/2020/5/20/no-harm-here-is-still-harm-there-looking-at-the-green-new-deal-from-the-global-south.

42 Ulrich Brand and Markus Wissen, *The Imperial Mode of Living. On the Exploitation of Human Beings and Nature in Global Capitalism*. London: Verso, 2020.

43 Giacomo D'Alisa, 'Circular Economy' in Kothari et al. (eds.), 2019.

Chapter 16

1 Dipak Gyawali, 'World Social Forum: Rethinking and Redefining Development Itself', *Economics and Technologies*, 22 February 2024, p. 5; Miriam Lang, Claus Konig and Ada-Charlotte Regelmann (eds.), *Alternatives in a World of Crisis*. Brussels: Rosa Luxemburg Stiftung, 2018.

2 Kirk Huffmann, 'Oceania's *Kastom Ekonomi*' in Ashish Kothari, Ariel Salleh, Arturo Escobar, Federico Demaria and Alberto Acosta (eds.), *Pluriverse: A Post-Development Dictionary*. New Delhi: Tulika and AuthorsUpFront, 2019, pp. 15–18.

3 Isabel Ortiz, 'The World Social Forum: The Counterweight to the World Economic Forum', *Other News: Voices against the Tide*, 23 February 2024: https://www.other-news.info/the-world-social-forum-the-counterweight-to-the-world-economic-forum/; Verdict of The Peoples Tribunal: People and Nature vs the UNFCCC, 7 November 2021: http://radicalecologicaldemocracy.wordpress.com.

4 Kothari *et al.*, 2019; Caroline Shenaz Hossein and Samuel Kwaku Bonsu. 'Situating the West African System of Collectivity: A Study of Susu Institutions in Ghana's Urban Centers', *Rethinking Marxism*, 35 (2023) 108–34.

5 Michel Pimbert, Nina Moeller, Brajesh Singh, Colin Anderson, 'Agroecology' in *Oxford Research Encyclopedia of Anthropology*, p. 9: https://doi.org/10.1093/acrefore/9780190854584.013.298.

6   Jason Hickel, 'The Problem with Saving the World', *Jacobin Magazine*, 8 August 2015.

7   Ariel Salleh, 'Editorial: Towards an Embodied Materialism', *Capitalism Nature Socialism*, 16 (2005) 9–14; John Clark, *Between Earth and Empire: From the Necrocene to the Beloved Community*. Oakland, CA: PM Press, 2019; Tim Hollo, *Living Democracy: An Ecological Manifesto for the End of the World as We Know It*. Sydney: New South Publishers, 2022; Matt York, *Love and Revolution: A Politics for the Deep Commons*. Manchester: Manchester University Press, 2023.

8   Pablo Solon, 'What Are Systemic Alternatives?' *Resilience*, 11 April 2019; https://www.resilience.org/stories/2019-04-11/what-are-systemic-alternatives/; 'Eco-Villages': http://gen.ecovillages.org; Simplicity Institute: http://simplicityinstitute.org/why-simplicity; *The Commoner*: http://www.commoner.org.

9   David Graeber, 'Bullshit Jobs: Direct Democracy and the End of Capitalism', *Novara Media*, 10 September 2020.

10 Giorgos Kallis, Susan Paulson, Giacomo D'Alisa and Federico Demaria, *The Case for Degrowth*. Cambridge: Polity Press, 2020; Vincent Liegey and Anitra Nelson, *Exploring Degrowth*. London: Pluto Press, 2020; Degrowth: http://www.degrowth.info.

11 Maria Mies, *Patriarchy and Accumulation on a World Scale*. London: Zed Books, 1987; Veronika Bennholdt-Thomsen and Maria Mies, *The Subsistence Perspective*. London: Zed Books, 1999; Ariel Salleh, 'Global Alternatives and the Meta-Industrial Class', in R. Albritton *et al.* (eds.), *New Socialisms: Futures beyond Globalization*. London: Routledge, 2004.

12 Karl Marx, *The Economic and Philosophic Manuscripts of 1844*. New York: International Publishers, 1964.

13 Via Campesina: https://viacampesina.org/en/; World Rainforest Movement: https://www.wrn.uy;

Fatimah Kelleher, 'Why the World Needs an African Ecofeminist Future', *African Arguments*, 12 March 2019; Sunaina Arya and Aakash Singh Rathore (eds.), *Dalit Feminist Theory: A Reader*. New Delhi: Routledge, 2019; Revolutionary Women of North East Syria, *Monthly Newsletter*, August 2022: Kongra-star.org; World March of Women: www.marchemondiale.org; 1000 Women for Peace Across the Globe: http://1000peacewomen.org; Code Pink: www.codepink.org.

14 Mary Graham, 'Place and Spirit: Spirit and Place', *EarthSong*, 2 (2014) 5–7; Anne Poelina, 'A Coalition of Hope! A regional governance approach to Indigenous Australian cultural wellbeing' in A. Campbell, M. Duffy and B. Edmonton (eds.), *Located Research, Regional Places, Transitions and Challenges*. New York: Springer, 2020.

15 Global Tapestry of Alternatives, 'Transformative Learning and Education', *Weaving Alternatives* August 2022: Global Tapestry of Alternatives: https://globaltapestryofalternatives.org/; Government of Bhutan, Gross National Happiness Survey: http://www.grossnationalhappiness.com/.

Government of Bhutan, Gross National Happiness Survey: http://www.grossnationalhappiness.com/.

16 Jackie Smith and Michael Goodhart, 'Introduction: Human Rights Globalization: How Local and Global Actions Institutionalize Human Rights', *Journal of Human Rights*, 23/2 (2024) 125–33.

17 Robert Fletcher *et al.*, 'A New Future for Conservation', *Progressive International*, 11 August 2020: http://progressive.international/blueprint/e6e09a90-dc09-410d-af87-5d3339ad4ed3-fletcher-et-al-a-new-future-for-conservation/en.

18 Ngakiya Camara and Kya Chen, 'Students Are Pushing US Colleges to Sever Ties With Military-Industrial Complex', *Truthout,* 7 Nov 2021: https://truthout.org/articles/students-are-pushing-us-colleges-to-sever-ties-with-military-industrial-complex/; Dissenters: https://wearedissenter.org/.

19 Kirkpatrick Sale, 'Principles of Bioregionalism' in J. Mander and E. Goldsmith (eds.), *The Case against the Global Economy*. San Francisco: Sierra Club, 1996; Vikalp Sangam: http://vikalpsangam.org; Juan Manuel Crespo, Shrishtee Bajpai, Ashish Kothari, 'Nation-states are destroying the world: Could 'bioregions' be the answer?', *Open Democracy*, 7 March 2022: https://www.opendemocracy.net/en/oureconomy/nation-states-are-destroying-the-world-could-bioregions-be-the-answer/.

20 Karin Amimoto Ingersoll, 'Sea Ontologies' in Kothari *et al.*, 2019, pp. 299–300.

Author Index

Abdelmalik, Philip 262
Acosta, Alberto 205, 206, 234, 235, 238, 246, 253, 260, 262, 263, 270
Adam, Barbara 254
Adams, W. 5, 232
Adler, David 259
Adorno, Theodor 62, 242
Agrawal, Anil 67, 243
Aguiton, Christophe 255
Ahmed, Nafeez 231
Altman, Jon 40, 237
Anderson, Colin 253, 268
Andreson, Steinar 244
Andrews, Peter 116, 249, 252
Angus, Ian 259
Anon 261
Antrobus, Peggy 124, 256, 260
Arcidiacono, S. 245
Arista, Noelani 266
Astyk, Sharon 247
Aviles Vazquez, K. 254

Baatjes, Britt 266
Badgley, Catherine J. 254
Bahro, Rudolf 19, 256
Baier, W. 259
Bainbridge, Alex 252
Bandara, Pri 263
Barad, Karen 213, 266
Barca, Stefania 169, 259
Barlow, Maude 53, 185, 186, 240
Barnosky, A. 265
Bar-On, Yinon 261
Barry, J. 5, 232

Barton, Mathew 200, 263
Bauer, S. 242
Baxter, Tim 264
Beck, Ulrich 64, 242
Becker, M. 227, 234, 255, 270
Behrens, William 65, 241
Belcher, Oliver 264
Bello, Walden 104, 189, 250
Benjamin, Medea 250
Bennett, Jane 266
Bennholdt-Thomsen, Veronika 73, 226, 244, 248, 269
Beresford, Quentin 237
Bertell, Rosalie 204, 265
Biermann, Frank 60–6, 68, 241–4, 263
Bigger, Patrick 264
Bird Rose, D. 5, 232
Blaser, Mario 218, 267
Bloom, Peter 205, 206
Boff, Leonardo 20, 235
Bond, Patrick 182, 241, 250, 260
Bonsu, Samuel Kwaku 268
Brahin, Philippe 265
Brand, Stewart 220, 241, 268
Brewer, Graham 267
Brockington, D. 5, 232
Bromley, Daniel 60, 63, 242
Broome, Neena Pathak 249, 257
Brundtland, Gro Harlem 51, 59, 65, 90, 161, 176, 247
Brunelle, D. 227, 234, 255, 270
Bruno, Kenny 236, 240
Bryant, Raymond 233, 241, 260, 264
Burmeier, Sandra 265

Burton, Bob   236
Busalacchi, Antonio   206, 265
Buscher, B.   5, 232
Butler, Simon   240, 259
Butler, Tom   263

Caffentzis, George   34, 35, 53, 237, 240
Caldecott, Leonie   49, 239
Calderon, Felipe   149, 256
Caldicott, Helen   44, 48, 238
Camara, Ngakiya   270
Canavan, Gerry   240
Canepa, E.   259
Carpenter, David   263
Carr, Susan   245
Carroll, William   242
Castells, Manuel   57, 241
Castree, Noel   5, 232
Cearreta, A.   265
Chandler, Michael   240
Changyong, Zhou   253
Chapin, F.   5, 207, 232, 243
Chase Dunn, C.   227, 234, 255, 270
Chen, Kya   270
Cheney, Joyce   48, 239
Chi, Lau Kin   253
Ching, Lim Li.   245
Chivers, Danny   260
Chossudovsky, Michael   237
Chretien, N.   245
Christopherson, Christine   236
Church, George   246
Chwalczyk, Franciszek   10, 254
Clark, Brett   244
Clark, John   6, 232, 233, 237, 261, 269
Clarke, Tony   53, 185, 186, 240
Colpani, Gianmaria   233
Comas, Julia Marti   173, 259
Commoner, Barry   245
Contreras, Lourdes   261
Cook, Alice   239
Cooper, Melinda   246
Corell, R.   5, 207, 232, 243
Cornera, E.   5, 232

Costanza, R.   5, 207, 232, 243
Crabtree, J.   267
Crawford, Kate   266
Criado Perez, Caroline   266
Crick, Francis   78, 245
Crutzen, Paul   4, 5, 205, 207, 231, 232, 232, 243, 243, 265
Cullinan, Cormac   234, 240
Cuomo, Chris   239

da Rimini, Francesca   202, 264
D'Alisa, Giacomo   268, 269
Dalla Costa, Mariarosa   237
Daly, Herman   135, 254
Danaher, Mike   239
Dartnell, Lewis   231
Davies, Nicolas   250
De la Cadena, Marisol   218, 267
De Sousa Santos, Boaventura   267
de Wit, C.   5, 207, 232, 243
Delheim, Judith   247
Della Porta, D.   227, 234, 255, 270
Demaria, Federico   234, 235, 238, 246, 253, 260, 262, 263, 269, 270
Demeritt, D.   5, 232
Di Fiore, Monica   13, 231, 246
Dickens, Peter   134, 254
Dion, Marc   262
Dire, Tladi   245
Dodson, Michael   31, 37, 237
Dodson Gray, Elizabeth   48, 239
Dowie, Mark   263
Dregger, Leila   261
Duffy, R.   5, 232
Duguay, F.   245
Dumble, Lynette   33, 34, 237

Easterly, William   180, 260
Edgeworth, M.   265
Elebi, Carolina Martinez   259
Elizalde, Antonio   73, 235, 244
Ellis, E.   265
Erickson, Jon   135, 254
Erik, Swyngedouw   5, 232
Ernstson, Henrik   5, 232

Escobar, Arturo xv, 11, 12, 218, 233–5, 238, 246, 253, 260, 262, 263, 267, 268, 270
Esteva, Gustavo 19, 234
Ewbank, Leigh 248

Fabry, V. 5, 207, 232, 243
Fair, Hannah 233
Falkenmark, M. 5, 207, 232, 243
Falkner, R. 242
Fanon, Frantz 267
Farley, Joshua 135, 254
Fauci, Anthony 85, 247
Federici, Silvia 35, 53, 135, 237, 240, 254
Felt, U. 5, 232
Figueroa-Helland, Leonardo 12, 172, 233, 234, 254, 259
Filippa, Lentzos 247, 262
Firstenberg, Arthur 265
Flanagan, Frances 265
Fletcher, Robert 270
Foley, J. 5, 207, 232, 243
Folke, C. 5, 68, 207, 232, 243
Foster, John Bellamy xvi, 6, 64, 92, 134, 167, 232, 242, 244, 248, 254, 258, 259, 264
Foucault, Michel xii, 11, 69, 244
Fourmile, Henrietta 35, 36, 237
Fox Keller, Evelyn 242
Francisco, Gigi 256
Freeland, Bill 237
Friday, George 256
Funtowicz, Silvio 72, 110, 115, 244, 246, 251
Furlong, Mark xii, xiv, 231

Galaz, V. 68, 243
Garcia, Edgardo 255
Gareau, Brian 259
Gautney, Heather 248
Georgescu-Roegen, Nicholas 134, 136, 254
Giddens, Anthony 64, 242
Gilbertson, Tamra 235
Gills, Barry 3, 231, 262, 265, 267

Glendinning, Chellis 50, 239
Goebel Noguchi, Mary 239
Golemis, H. 259
Goodhart, Michael 270
Goodman, James 202, 241, 247, 250, 262, 264, 265, 267
Goodwin, B. 245
Gorshkov, Victor 112, 251
Gowdy, John 136, 254
Graeber, David 226, 269
Graham, Mary 227, 269
Graham, Whitney 238
Green, Jim 44, 238
Greer, Jed 240
Griffin, Susan 47, 239
Grinevald, J. 265
Grinevald, Jacques 4, 5, 205, 231, 232, 243, 265
Grusin, Richard 232
Gupta, Aarti 244
Gyawali, Dipak 223, 268

Haeringer, Nicolas 255
Haff, P. 265
Hagerdorn, Kurt 243
Hajer, Martin 247
Hakusui, Shigeaki 203, 264
Hansen, J. 5, 207, 232, 243
Harari, Yuval 199, 263
Haraway, Donna 16, 212, 232, 266
Harcourt, Wendy 233
Harding, Sandra 233
Hardt, Michael 57, 241
Hart, Amalia 264
Hartmann, Betsy 236
Harvey, David 178, 260
Harvey, Fiona 250
Hawken, Paul 64, 242
Hay, Peter 233
Heenan, Natasha 167, 258
Heinberg, Richard 248
Henderson-Sellers, Ann 111, 251
Herrera, Remy 151, 253
Hertsgaard, Mark 263
Hesslerova, Petra 112, 248, 251

Heyman, Ruth   238
Hickel, Jason   xv, 176, 178, 220, 253, 260, 268, 269
Hinkson, John   261
Hinkson, Melinda   236
Hinman, Pip   238
Hogan, Anna   261
Hollo, Tim   269
Hooks, Bell   233
Hopenhyan, Martin   73, 235, 244
Horkheimer, Max   62, 242
Hornborg, Alf   5, 214, 232, 236, 267
Hossay, Patrick   23–6, 234
Hosseini, Hamed   232, 256, 262, 265, 267
Howey, Kirsty   238
Huang, Y.   245
Hubbard, Ruth   245
Huffmann, Kirk   268
Hughes, T.   5, 207, 232, 243
Hynes, Pat   33

Illich, Ivan   19, 234, 251
Ingersoll, Karin Amimoto   229, 270
Isla, Ana   136, 248, 252
Ivar do Sul, J.   265

Jackson, Tim   158, 257
Jahi Chappell, M.   254
James, Annie   249, 257
Jasanoff, Sheila   72, 76, 113, 244, 245, 251
Jeandel, C.   265
Johnsson-Latham, Gerd   248
Joler, Vladan   266
Jones, Lynne   239
Joy, K. J.   234
Jun, Wen Tie   253
Jungcurt, S.   66, 68, 243

Kallis, Giorgos   220, 268, 269
Karamoskos, Peter   40, 238
Karides, M.   227, 234, 255, 270
Karkal, Malini   236
Karlberg, L.   5, 207, 232, 243
Kelleher, Fatomah   269
Kelly, Petra   49, 239

Kennan, George   59, 241
Kennelly, Cara   264
Kiley, Sam   237
Kinoshita, Chigaya   238
Kirk, Gwyn   239
Kite, Suzanne   266
Klarr, Lisa   240
Klein, Naomi   84, 192, 221, 262
Klein, Renate   246, 262
Koeppel, Barbara   263
Kohutiar, Juraj   252
Kolinjivadi, Vijay   157, 220, 257, 268
Konig, Claus   268
Konik, Inge   235
Kothari, Ashish   157, 220, 234, 235, 238, 246, 253, 257, 260, 262, 263, 268, 270
Koushik, Jade Reness   234
Kravcik, Michal   109, 116, 184, 251, 252
Kroger, Markus   13, 233, 241
Kuhn, Thomas   133, 251
Kurzweil, Ray   267

Lai, H.   267
Lambin, E.   5, 207, 232, 243
Lang, Chris   248, 256
Lang, Miriam   261, 268
Langton, Marcia   29, 38, 236
Lash, Scott   64, 242
Latour, Bruno   7, 57, 212, 216, 232, 241, 266
Lau, Kin Chi   151, 253
Lazaris, A   245
Le Grange, Lesley   235
Leahy, Terry   235
Ledgar, Richard   236
Leinfelder, R.   265
Leland, Stephanie   49, 239
Lenton, T.   5, 207, 232, 243
Lerner, Stephen   154, 256
Levidow, Les   245
Levin, S.   83, 242
Lewis, Jason   211, 266, 267
Li, Guanqui   253
Li, Lim Lin   245
Liverman, D.   5, 207, 232, 243

Lohmann, Larry   109, 251
Lopez, Belinda   254
Lovins, Amory   64, 242
Lovins, Hunter   64, 242
Lucas, Caroline   257
Luke, Timothy   69, 72, 244
Luxemburg, Rosa   42, 59, 235, 241

Macalister, Terry   250
MacGregor, Sherilyn   239
MacLellan, Nic   240
Mahnkopf, Birgit   259
Maina-Okori, Mumbi   234
Makarieva, Anastassia   112, 251
Malm, Andreas   5, 232
Marshall, Jonathan   202, 264
Martin, David   86
Martinez-Alier, Joan   239, 259
Marx, Karl   xii, xiii, xvi, xvii, 6, 15, 73, 92, 104, 176, 207, 226, 269
Masha, Borak   264
Mawudeku, Abla   262
Max-Neef, Manfred   73, 235, 244
May, Robert   83, 242
McAlpine, Clive et al.   251
McDonagh, Sean   19, 235
McMullen, Mathew   233
McNeill, John   4, 5, 205, 231, 232, 243, 265
Meadows, Dennis   65, 241
Meadows, Donella   65, 241
Mellor, Mary   231, 235, 244, 253
Mendes, Margarida   264
Merchant, Carolyn   48, 239
Mertes, Tom   255
Michael, Shellenburger   220
Mies, Maria   xvi, 18, 50, 73, 226, 235, 239, 244, 248, 254, 269
Mignolo, Walter   11, 233
Milne, Christine   258
Milo, Ron   261
Moeller, Nina   253, 268
Moghtader, E.   254
Mol, Arthur   90, 264, 269
Mollison, Bill   19, 235, 257

Moore, Jason   xvi, 5, 232
Morales, Evo   95, 96, 172, 248
Morita, Keitaro   238
Mpofu, Elizabeth   255
Muchhala, Bhumika   173, 259
Murphy, Patrick   253

Ndlovu-Gatsheni, Sabelo   218, 267
Neal, C.   265
Negri, Antonio   57, 241
Neimark, Ben   264
Nelkin, Dorothy   239
Neubauer, Deane   191, 262
Neves, K.   5, 232
Newell, P.   5, 232
Newman, Stuart   76, 245
Nicholls, Zebedee   264
Nicklasson, Elena   238
Nilsson, M.   68, 69, 243, 244
Noone, K.   5, 207, 232, 243
Nordhaus, Ted   220
Norris, Duane   116, 249, 252
Ntloketse, Ruth   262
Nykvist, B.   5, 207, 232, 243

Obe, Mitsuri   238
Ocasio-Cortez, Alexandria   167, 170, 258
O'Connor, James   xvi, 178, 260
Odada, E.   265
Olsson, P.   68, 243
Oreskes, N.   265
Orsini, A.   242
Ortiz, Isabel   266
Ostrom, Elinor   68, 69, 243

Padmanabhan, M.   66, 68, 243
Palmesino, J.   265
Partzsch, L.   244
Pascoe, Bruce   39, 237
Pattberg, Phillip   60, 241, 263
Paulson, Susan   269
Pearce, Daryl   31
Pearce, Fred   249
Pechawis, Archer   266
Peet, Richard   233

Pellizzoni, Luigi   5, 19, 232, 253
Perfecto, I.   254
Persson, A.   5, 68, 69, 207, 232, 243, 244
Phelps, Bob   237
Phillips, Rob   261
Pimbert, Michel   253, 268
Pitron, Guillaume   267
Pleyers, Geoffrey   234
Plumer, Brad   264
Poelina, Anne   227, 269
Pokorný, Jan   112, 248, 251
Pope Francis   235
Portal, Aurora   261
Posey, Darrell   237
Prakash, Madhu Suri   19, 234
Price, S.   265
Prigogine, Ilya   251
Pulvers, Roger   45, 238

Quijano, Annibal   33
Quintero, E.   254

Raghu, Pratik   12, 172, 233, 234, 254, 259
Ramose, Mogobe   19, 235
Randers, Jorgen   65, 241
Ravetz, Jerome   72, 110, 244, 246, 251
Redshaw, Megan   247
Rees, William   254
Reeves, Guy   247, 262
Regelmann, Ada-Charlotte   268
Regenvanu, Ralph   244, 249
Regis, Ed   246
Reilly, Bethany   253
Renner, Michael   238
Revkin, A.   265
Richardson, K.   5, 207, 232, 243
Rigby, Kate   5, 232
Ripl, Wilhelm   134, 249, 251
Rivero, Rosa   261
Robbins, Paul   5, 232, 233, 241
Robin, L.   5, 232
Rocheleau, Dianne   20, 235
Rockström, Johann   5, 207, 232, 243
Rodhe, H.   5, 207, 232, 243

Rommetveit, Kurt   110, 115, 251
Rönnskog, A.   265
Rosa, Eugene   64, 90, 242, 247, 251
Rose, Deborah   232, 254
Rosewarne, Stuart   247
Ross, A.   5, 232
Rudy, Alan   259
Russell, George   256
Ryall, Julian   240

Sachs, Wolfgang   19, 234
Sahlins, Marshall   237
Saito, Kohei   xvii, 6, 232
Sale, Kirkpatrick   19, 136, 235, 254, 270
Sales, Louise   250, 264
Salleh, Anna   245
Salleh, Ariel   234, 235, 238, 246, 253, 260, 262, 263, 270
Saltelli, Andrea   13, 231, 246
Samulon, A.   254
Sanchanta, Mariko   238
Sanders, Bernie   167
Sapinski, Jean-Paul   242
Satgar, Vishwas   262
Saunders, P.   245
Scasserra, Sofia   259
Scheffer, M.   5, 207, 232, 243
Schellnhuber, H.   5, 207, 232, 243
Schlosberg, D.   5, 232
Schmidheiny, Stephan   236, 239, 241
Schneider, Reto   265
Schrodinger, Erwin   75, 245
Schwab, Klaus   191, 192, 216, 262, 267
Serrano, Isagani   255
Shenaz Hossein, Caroline   268
Shilliam, Robbie   233
Shiv, Visvanathan   83, 240
Shiva, Vandana   xv, 20–6, 50, 53, 136, 225, 231, 233–6, 239, 242, 244, 245, 252, 258, 261, 262
Shorrock, Tim   262
Siebenhuner, Bernd   244
Singh, Brajesh   253, 268
Singh Rathore, Aakash   269

Sitrin, Marina   255
Sklair, Leslie   242
Smith, Jackie   227, 234, 255, 270
Snyder, P.   5, 207, 232, 243
Solon, Pablo   19, 96, 118, 235, 252, 269
Song, Xin   253
Song, Yiching   253
Sonnenfeld, David   90, 264, 269
Sörlin, S.   5, 207, 232, 243
Sorlin, Z.   5, 232
Spash, Clive   253, 257
Spitzner, Meike   240
Steele, Mark   263
Steffen, Wil   4, 5, 205, 207, 231, 232, 232, 243, 243, 265
Steisslinger, M.   259
Stevens, Tina   245
Steyn, Elizabeth   265
Stirling, Andrew   244
Strand, Roger   110, 115, 251
Strom, Timothy   263
Sturman, Anna   167, 258
Sugihara, G.   83, 242
Summerhayes, C.   265
Sunaina, Arya   269
Svampa, Maristella   241
Svedin, U.   5, 207, 232, 243

Tabuchi, Hiroko   264
Tamale, Sylvia   234
Taylor, Maureen   256
Temper, Leah   234
Thomas-Slayter, Barbara   20, 235
Thunberg, Greta   11, 234
Tienhaara, K   242
Tokar, Brian   235
Toulmin, Stephen   113, 251
Tout, Dan   231
Trainer, Ted   19, 169, 235, 243, 259
Trueman, Charlotte   264
Tsui, Sit   19, 130, 253
Tuhiwai Smith, Linda   267
Tulus, Elna   253
Tuzcu, Pinar   266

Ulrich, Brand   15, 80, 221

Van Beers, Britta   64, 246
van der Leeuw, S.   5, 207, 232, 243
Van der Pijl, Kees   71, 244
Varoufakis, Yanis   169, 192, 262
Venter, Craig   75, 245
Vernooy, Ronnie   253
Vidas, D.   265
Von der Leyen, Ursula   16
Vu, Ryan   240
Vyas, H. M.   234

Wackernagel, Mathias   254
Wald, George   245
Walker, B.   5, 207, 232, 243
Wallace, Rob   189, 261
Walsh, Toby   267
Wan Ho, M.   245
Wangari, Esther   20, 235
Wargan, Pawel   259
Waring, Marilyn   166, 244, 258
Waters, C.   265
Watson, James   78, 245
Watts, Michael   233
West, P.   5, 232
Whaanga, Hemi   217, 267
Whitaker, Chico   154, 256
White, Damian   259
Whitehead, M.   5, 232
Whyte, Kyle   19, 235
Williams, M.   265
Williamson, Ben   261
Wilson, Alexandria   234
Wing, S.   265
Winiecki, Martin   261
Wissen, Markus   15, 80, 221
Wolfe, A.   265
Worth, Jess   260
Wynne, B.   5, 232
Wynne, Brian   63, 232, 242, 246, 247

York, Matt   269
York, Richard   64, 90, 242, 244, 247, 251

Young, Oran 60, 242
Yunkaporta, Tyson xv, 238

Zakem, E. 254
Zalasiewicz, Jan 265

Zhou, J. 245
Ziegler, R. 244
Zowghi, Didar 202, 264
Zuboff, Shoshana xv, 195, 210, 262, 266

Subject Index

1/0 imaginary xv, xvi, 7, 12, 14, 16, 35, 40, 57, 75, 109, 117, 120, 135, 192, 195, 206, 207, 211, 216, 228
2030 Agenda ix, 175–87

Abrahamic traditions 119
Agrarian South Network 3, 231
agritech 138, 139
agroecology 127–38, 170, 253, 254, 268
Amazon, corporation 168, 194, 195, 204, 266
Amazon, jungle 13, 107, 233, 256
American Association for the Advancement of Science (AAAS) 236
Androcene/androcentric vii, xi–xvii, 4–17, 20, 33, 39, 40, 48, 52, 55, 56, 62, 73, 75, 90, 91, 110, 116, 120, 123, 145, 146, 172, 174, 217, 223, 228
andro-science 83, 186, 193, 206, 225
Another Future is Possible! ix, 4, 15, 35, 125, 141–56, 224–9
Anthropocene 2–4
Appeal on 5G to UN, WHO, EU, and Nation States 263
artificial intelligence (AI) 188, 191, 193, 195, 209, 211, 212, 215, 265–7
Asia-Pacific Forum on Women, Law and Development 54, 240
Australian Conservation Foundation (ACF) 32, 164, 165
Australian Council of Trade Unions (ACTU) 164, 258
Australian Cyber Security Centre (ACSC) 195, 263
Australian Green Infrastructure Council (AGIC) 165
Australian Naval Nuclear Power Safety (Transitional Provisions) Bill 240

Beijing 1995, Women's Platform of Action 33, 124, 173, 255
Beijing Academy of Science 132, 253
Beyond Zero Emissions 94, 248, 251
Big Data 138, 190, 192, 207, 215, 216, 224, 262, 264
Big Pharma 24, 27, 34, 35, 76, 82, 84, 122, 132, 138, 190, 192, 207, 224
biocolonialism 7, 35, 142, 153
biopiracy 21, 35, 67, 83, 122, 152, 237
biopolitics xvii, 191, 232, 245
bioregionalism 29, 41, 130, 228, 235, 270
biosecurity 51
Breakthrough Institute 6, 209, 219, 268
buen vivir xv, 15, 19, 89–106, 123, 130, 152, 155, 180, 187, 224, 225, 235
Business Action for Sustainable Development (BASD) 141, 142, 256

capacity building 32, 34, 52, 62, 96, 102, 125, 133, 136, 143, 181, 186
carbon fetishism 107–18, 200
Centers for Disease Control (CDC) 186, 198, 247
Chavistas 226
Chernobyl 42, 50, 238
Chinese women farmers 131, 132
circular economy 168, 199, 220, 221, 224, 268

CITES, International Trade in Endangered Species 66
climate finance 94, 180–3
Clustered Regularly Interspaced Short Palindromic Repeats (CRISPR) 84, 87
Code Pink 50, 54, 122, 269
coexistence 81
commoning 4, 125, 130, 148, 152, 153, 155, 169, 225, 227
Commonwealth Scientific and Industrial Research Organization (CSIRO) 83, 245
complexity 4, 24, 51, 63, 64, 66–9, 75, 78, 111, 136, 216, 231, 234, 244, 246, 251
contradiction xv, xvi, 7, 16, 40, 46, 75, 98, 124, 147, 164, 178–80, 184, 186, 209, 210, 218, 243
Convention on Biological Diversity 25, 29, 37, 51, 59, 79, 107, 113, 158, 202
corporate capture 3, 12, 32, 121, 192, 265
Council of the European Union 198, 263
Countercurrents Collective 261
Covid-19 xvii, 3, 83, 86, 87, 189, 190, 200, 216, 227, 247, 261–3
Cultural Survival 38
cyberfuture 25, 57, 135, 195, 203, 209, 213, 216, 217, 262, 263, 266

Dalit women 226, 269
data 30, 32, 38, 45, 62, 63, 84, 112, 115, 133, 138, 151, 162, 168, 177, 178, 190–200, 202, 207, 262, 264, 266
data sovereignty 209, 213, 215, 216, 224
Davos 4, 59, 125, 177, 192, 199
decolonial xi, xv, xvi, xvii, 4, 10–12, 15, 20, 26, 30, 52, 96, 97, 123–5, 130, 173, 181, 185, 217, 219, 225, 226, 233, 259
decoupling 64, 219, 220
deep ecology xii, 16, 20, 239
Defense Advanced Research Projects Agency (DARPA), US 222, 264
degrowth 4, 123, 125, 130, 169, 187, 220, 226, 232, 239, 261, 269

Democracy in Europe Movement (DiEM25) 168, 169, 220, 259
Democratic Socialists of America (DSA) 170, 171, 220, 259
denial 11, 13, 16, 40, 44–6, 75, 163
digital xiii, xv, 3, 10, 16, 26, 30, 37, 62–4, 72, 93, 118, 123, 133, 138, 150, 151, 168, 189–95, 199–201, 203, 205, 207–19, 259, 262, 263, 266
dissociation xvi, 7, 10, 19, 40, 74, 207, 225, 229
dualisms vi, xii, xiii, 7, 8, 11, 16, 119, 121, 134, 157, 176, 214

Earth Summit 3, 29, 30, 39, 51, 52, 59, 108, 132, 141, 146, 164, 177
ecocentrism xvi, 39, 228, 229
ecofeminism xi–xvii, 1–12, 20–7, 33, 46, 49–52, 54, 55, 70, 84, 93, 114, 119, 136, 160, 166, 169, 173, 193, 204, 210, 217, 226, 229, 230, 232, 233, 235, 236, 238, 239, 244, 252–4, 257, 259, 265, 269
ecomodernism xv, 3, 6, 15, 16, 27, 51, 54, 62, 63, 89, 90, 100, 105, 123, 134, 139, 157, 158, 190, 202, 213, 216, 220, 221, 228
ecosocialism vi, vii, xv, xvii, 6, 15, 16, 20, 50, 72, 101, 104, 205, 226, 229, 233, 241, 259, 260, 265
eco-sufficiency viii, 14, 19, 97, 102, 122, 136, 233, 235, 240, 244, 248, 250–4, 258, 260
embodied debt 56, 58, 91, 105, 129, 143, 148, 150, 163, 172, 228
embodied materialism xiv, 16, 62, 121, 220, 241, 269
Enlightenment narrative 9, 12, 24, 62, 68, 73, 119, 242
Environmental Health Trust 263
Environmental Justice Atlas (EJAtlas) 234
Environmental Justice Organizations, Liabilities and Trade (EJOLT) 151, 256

Erosion, Technology, Concentration (ETC) 142, 190, 193, 201, 202, 246, 250, 252, 256, 262, 264
Eurocentrism xiii, xvii, 6, 7, 9, 15, 30, 37, 52, 91, 92, 94, 97, 99, 113, 120, 121, 123, 139, 145, 149, 152, 168, 189, 211, 214, 218, 224, 226
exchange, thermodynamic 15, 74, 104, 122, 160, 166, 169, 225
exchange, unequal 5, 14, 36, 64, 70, 145, 152, 267
Extinction Rebellion xv, 1, 15, 155, 167, 228, 233, 257
extractivism xii, 2, 13, 29, 30, 58, 71, 121, 129, 150, 168, 172, 178, 187, 235, 241, 265

Federal Communications Commission (FCC), US 196, 198, 203
First Nations National Constitutional Convention, Uluru 238
food sovereignty 22, 23, 31, 100, 104, 115, 122, 125, 127–40, 142, 148, 155, 166, 170, 182, 224, 226, 253, 254
Fourth Industrial Revolution (4IR) xv, 2, 33, 111, 189–208, 211, 213, 214, 216, 217, 262, 267
Friends of the Earth (FOE) 47, 54, 109, 167, 236, 239, 250
Fukushima 43–56, 238, 240

Gabaewul, South Korea 226
Gandhi 19, 20
Gates Foundation 26, 27, 132, 136, 190, 210, 233
Geek Girl Academy 217, 267
Gender and Climate Change (GenderCC), Berlin 240
GeneEthics Network 88
Gene Technology Act 2000, No. 169, AU 78
genetic engineering 16, 36, 37, 50, 53, 64, 75–88, 121, 189, 190, 192, 215, 246
German Advisory Council on Global Change 65

Global Alliance for Vaccines & Immunization (GAVI) 85
Global Environment Facility (GEF) 59, 97, 100, 159, 249, 251
Global Redesign Initiative, Davos 177, 192, 195, 260
Global Tapestry of Alternatives 227, 269
Global University for Sustainability 15, 129, 225, 228
GM Watch 246
Great Chain of Being 8
Green New Deals 157–74
Gross National Happiness Survey, Bhutan 269

Hawaii fishers 229
hierarchy xxvii, 8, 12, 13, 46, 120, 129, 146, 152, 194, 221
High Frequency Active Auroral Research Program (HAARP), US 204
holding labour 141, 228, 229

IBON International, Manila 94, 139, 248, 249, 252
ILO Convention 169 83
imperial mode of living 58, 72, 221, 241, 268
Indigenous Regional Agreements 29–33
Intergovernmental Panel on Climate Change (IPCC) 72, 90, 96, 110, 111, 113, 115, 117, 155, 170, 181, 204, 251
International Atomic Energy Agency (IEA) 44, 45, 50, 55, 247
International Campaign to Abolish Nuclear Weapons (ICAN) 56
International Committee on Non-Ionizing Radiation Protection (ICNIRP) 199
International Conference on Organic Agriculture and Food Security 21, 36, 254, 257
International Convention on Civil and Political Rights 37
International Convention on Large Dams 22

International Institute for Sustainable
    Development (IISD) 144, 152,
    255, 256
International Labor Organization
    (ILO) 163
International Monetary Fund (IMF) 24,
    26, 34, 35, 105, 142, 237
International Trade Union Confederation
    (ITUC) 92, 93, 147, 148, 248
internet of bodies 189
internet of things 189, 191, 199, 200

Just Net Coalition 262
Just Transition Fund 168, 170, 259

Kalpavriksh 227
Kari-Oca Declaration and Indigenous
    Peoples Earth Charter 246
Košice Civic Protocol on Water 101,
    249, 251
Kyoto Protocol 25, 31, 90, 113, 162,
    248, 252

labour ix, xv, xvii, 7, 10, 13, 14, 20, 25,
    26, 34, 35, 45, 55, 59, 68, 73, 92, 94,
    104, 113–15, 120–2, 128, 129, 132,
    134–7, 145–8, 152, 157, 162, 164, 167,
    169–71, 176, 193, 220, 221, 226, 228,
    229, 246, 256
libidinal rift xi, xvi, 6, 10, 16
Little Donkey Farm Green Ground
    EcoTech Centre, Beijing 129, 253
livelihoods 19–28, 34, 39, 78, 87, 93, 122,
    134, 159, 180, 184, 202, 227

Manushi Collective, India 48, 239
Marxism xii, xiii, xvi, xvii, 6, 15, 62, 73,
    92, 105, 134, 167, 169, 176, 178, 202,
    207, 227, 232, 233, 236, 248, 254, 259,
    264, 268, 269
materialism xvi, 16, 121, 240, 241,
    259, 269
M-CAM, US 36, 247
metabolic rift xvi, 6, 16, 72, 92, 102, 136,
    169, 202, 203, 244, 255, 259, 264

metabolic value xv, xvi, 91, 100, 102,
    150, 152, 157, 160, 166, 169, 184, 244,
    249, 255, 259
meta-industrial class xv, xvi, 101, 104,
    122, 134, 136–9, 141, 148, 149, 152,
    164, 193, 221, 234, 255, 269
methodological forcing 63, 110, 115
Millennium Development Goals
    (MDGs) 124, 147, 149, 176, 260
modernity 124, 147, 149, 176, 260
Mujeres 53, 244, 255

nanotechnology 12, 123, 143, 191,
    193, 219
National Institutes for Health (NIH), US 79
National Organization of Women (NOW),
    US 240
needs 3, 47, 54, 55, 100, 127, 128, 131, 133,
    136, 141, 148, 151, 171, 175, 181, 185,
    192, 193, 205, 213, 225, 226, 263, 269
New Economics Foundation (NEF) 144,
    147, 256, 257
non-ionizing, Electromagnetic Fields
    (EMF) 9, 62, 196, 197, 203, 204, 206,
    219, 263, 267
nuclear risks xv, 30, 32, 43–56, 63, 110,
    112, 118, 121, 134, 161, 168, 182, 197,
    199, 206, 218, 220, 236, 239, 240, 263
Nuffield Council on BioEthics 87, 247

Occupy movement 142, 146, 153–5,
    255, 256
Oceania, South Pacific 30, 36, 54–6, 72,
    107, 118, 139, 237, 239, 240
Oceania Radiofrequency Scientific
    Advisory Association (ORSAA) 263
Office of the Gene Technology Regulator
    (OGTR) 76, 77, 79, 80, 84, 87, 245
ontology 7, 11, 30, 223–9
othering 7, 10, 223

patriarchal-colonial-capitalism xii, 3, 4,
    6, 9–17, 20, 21, 26, 37, 48, 53, 58, 73,
    75, 83, 93, 99, 107, 110, 117, 120, 123,
    125, 129, 135, 143, 145, 151, 173, 174,

176, 182, 189, 206, 208, 216, 220, 223, 226, 229
Peace-Women-Across-the-Globe   269
peoples science   103, 115–17, 131, 138, 139, 142, 225, 227, 246, 260
Peoples Tribunal: People and Nature vs UNFCCC   224, 268
Peoples World Conference on Climate Change (CMPCC), Bolivia   94–100, 248
petro-farming   26, 142, 168, 224, 231
Pluriverse   15, 219, 233–5, 238, 246, 253, 260, 262, 268
precautionary principle   22, 36, 79, 84, 102, 120, 125, 137, 141, 198, 214, 246, 250, 252
Progressive International   192, 227, 270

Radical Ecological Democracy   37, 234, 268
REDD Scheme   114, 134, 154, 159, 248, 250
reductionism   71, 244
regenerative models   12, 124, 137–8, 247, 252
reproduction   xii, xvii, 6, 7, 13, 34, 46, 78, 120, 148, 160, 169, 172, 201, 225, 237, 245, 247, 262
Rising Tide movement   270
Rockefeller Foundation   26
Rojava women   226
Rosa Luxemburg Foundation   4, 20, 59, 125, 157, 235, 241, 257, 259, 268

Shibokusa women   49
Simplicity Institute   269
Social Ecology   20
South African Climate Justice Charter   114, 262
South Peoples Historical, Social-Ecological Debt Creditors Alliance   258
space   xiii, 12, 14, 62, 87, 96, 107, 200, 203–6, 209, 213, 229, 263, 265

Stockholm International Peace Research Institute (SIPRI)   56, 240
Subaltern Studies   11
subsidiarity   4, 103, 114, 125, 139, 221
subsistence ethic   20, 22, 39, 90, 94, 96, 105, 114, 125, 130, 131, 133, 173, 202, 226, 227, 244, 248, 253, 254, 269
surveillance   190, 192, 194, 210, 224, 262, 266
Sustainable Development Goals (SDG)   108, 124, 175–86
Sustainable Europe Research Institute (SERI)   258
swaraj   15, 19, 22, 125, 224, 234
synthetic biology   11, 83, 84, 87, 88, 143, 246

Tebtebba Foundation   241
terra nullius   xiv, xv, 11, 16, 29–42, 173, 206, 218, 224, 227, 236
Thematic Social Forum, Dialogue Platform   142, 256
Third World Network (TWN)   249, 250
Trade Related Intellectual Property Rights (TRIPS)   35
transversal praxis   xiv, xvii, 4, 14, 225, 227, 232

ubuntu   15, 19, 125, 224, 235
uncertainty principle   57–72, 79, 111, 114, 116, 199
UN Commission on the Status of Women   146, 256
UN Committee on the Elimination of Discrimination Against Women (CEDAW)   54, 173
UN Consultative Group on International Agricultural Research (CGIAR)   36, 82
UN Convention on Biological Diversity   29, 36, 51, 59, 70, 107, 113, 132
UNCSD-Rio+20, The Future We Want   141, 143, 150, 177, 255
UN Declaration of the Rights of Indigenous Peoples   155

UN Environment Programme (UNEP)   58, 59, 62, 71, 100, 123, 131, 132, 143, 146, 147, 149, 150, 152, 153, 158–60, 163, 220, 252, 255–7, 268
UN Food and Agriculture Organization (FAO)   36, 82, 159, 236, 254
UN Framework Convention on Climate Change (UNFCCC)   93, 132
Universal Declaration for the Rights of Mother Earth   53, 95, 152, 153, 172, 240
UN Panel on Digital Cooperation   209, 210, 266
UN Women's MOU with Blackrock Bank   210, 266
UN Year of Indigenous Languages   215, 267
US Agency for International Development Aid (USAID)   85

Vaccine Adverse Event Reporting System (VAERS), CDC, US   86
Via Campesina   60, 91, 102, 125, 130, 142, 185, 226, 247, 249, 252, 255, 261, 269
Vikalp Sangam   19, 270

water   7, 13, 23, 24, 56, 79, 101, 107–18, 154, 184–7, 206, 236, 244, 248, 249, 252
Women and Life-on-Earth   48, 54, 121, 240
Women Working for a Nuclear Free and Independent Pacific   50
WoMin, Africa   122, 226
World Bank   11, 24, 34, 59, 71, 92, 122, 123, 135, 144, 159, 179, 210, 237
World Economic Forum (WEF), Davos   192, 196, 199, 205, 216, 267
World March of Women   18, 50, 145, 226, 261, 269
World Rainforest Movement (WRM), Uruguay   108, 226, 250, 269
World Social Forum   19, 95, 124, 142, 146, 152, 153, 227, 228, 232, 256, 268
World Trade Organization (WTO)   21, 24, 29, 36, 59–61, 79, 87, 97, 99, 105, 113, 150
Worldwatch Institute   25, 92, 144, 161, 163, 164, 256–8
World Wildlife Fund (WWF)   38, 149, 163